Springer Series in

ADVANCED MICROELECTRONICS 3

Springer

Berlin
Heidelberg
New York
Barcelona
Hong Kong
London
Milan
Paris
Singapore
Tokyo

Springer Series in

ADVANCED MICROELECTRONICS

The Springer Series in Advanced Microelectronics provides systematic information on all the topics relevant for the design, processing, and manufacturing of microelectronic devices. The books, each prepared by leading researchers or engineers in their fields, cover the basic and advanced aspects of topics such as wafer processing, materials, device design, device technologies, circuit design, VLSI implementation, and subsystem technology. The series forms a bridge between physics and engineering and the volumes will appeal to practicing engineers as well as research scientists.

1 **Cellular Neural Networks**
Chaos, Complexity and VLSI Processing
By G. Manganaro, P. Arena, and L. Fortuna

2 **Technology of Integrated Circuits**
By D. Widmann, H. Mader, and H. Friedrich

3 **Ferroelectric Memories**
By J. F. Scott

J. F. Scott

Ferroelectric Memories

With 170 Figures

Springer

Professor James F. Scott
Centre for Ferroics
Earth Sciences Dept.
Cambridge University
Downing Street
Cambridge CB2 3EQ
England

Series Editors:

Dr. Kiyoo Itoh
Hitachi Ltd., Central Research Laboratory
1-280 Higashi-Koigakubo
Kokubunji-shi
Tokyo 185-8601
Japan

Professor Takayasu Sakurai
Center for Collaborative Research
University of Tokyo
7-22-1 Roppongi, Minato-ku,
Tokyo 106-8558
Japan

ISSN 1437-0387
ISBN 3-540-66387-8 Springer-Verlag Berlin Heidelberg New York

Library of Congress Cataloging-in-Publication Data

Scott, J.F. (James Floyd), 1942-
Ferroelectric Memories / J. F. Scott.
p. cm. – (Advanced microelectronics ; 3)
Includes bibliographical references and index.
ISBN 3540663878 (alk. paper)
1. Ferroelectric storage cells. I. Title. II. Series. III. Springer series in advanced microelectronics ; 3.

TK7895.M4 S36 2000
621.39'732–dc21 99-056368

Typesetting: Data conversion by Springer-Verlag
Cover design: *design & production* GmbH, Heidelberg

Printed on acid-free paper SPIN 10741496 57/3144/xo 5 4 3 2 1 0

For my wife, Galya

Preface

The photograph opposite was taken by me in November 1970 in front of the old Department of Natural Philosophy on Drummond Street at The University of Edinburgh. The occasion was historic in that it was the first meeting of Dr. K. Alex Mueller (left) with Prof. William Cochran (centre) and Prof. Roger A. Cowley (right). It was my pleasure to spend that year with Cowley and Cochran (I shared an office with Bill). During the year I learned why certain crystals were considered ferroelectric at room temperature in Scotland but not in the USA. The buildings were unheated, and our basement laboratory hovered around 10°C summer and winter!

In the present book I try to combine the present knowledge of device physics for ferroelectric thin-film memories with information about circuit design, materials deposition and characterization, and testing protocols. It is my hope that this will serve as a useful reference for engineers and device physicists in industry and in government laboratories, and as a book that will assist the teaching of this topic in senior or graduate-level engineering and applied physics university classes.

With this book, I would like to honour the four people who most influenced directions of modern research on ferroelectricity – the three pictured in the photograph, and Dr. Helen Megaw, whose early role in the study of ferroelectric crystals at Cambridge is one I seem to have inherited.

Special thanks are also due to my colleagues of 1984–1992 at Symetrix, Carlos Araujo and Larry McMillan, without whom this work would never have begun.

Cambridge
March 2000

James F. Scott

Contents

Glossary of Symbols

α	Activation field for ferroelectric switching; critical exponent characterizing specific heat divergence slightly above T_0
α'	Critical exponent characterizing specific heat divergence slightly below T_0
A	Lateral area of memory cell
a	Lattice constant or jump distance for electrons in a ferroelectric semiconductor with hopping conduction
a'	Accommodation length (region of charge diffusion near an electrode)
a_0, b_0	Lateral dimensions of capacitor surface, with $a_0 < b_0$
ANNNI	Axial Next Nearest Neighbor Ising model
β	Activation energy for electrical conduction; critical exponent characterizing $P(T)$ near T_C
BMF	Barium Magnesium Fluoride $BaMgF_4$
BST	Barium Strontium Titanate $Ba_xSr_{1-x}TiO_3$
C	Capacitance
CMOS	Complementary Metal-Oxide Semiconductor
χ	Semiconductor electron affinity; electrical permittivity
C_V	Specific heat
CVD	Chemical Vapour Deposition
δ	Characteristic length in dP/dz at ferroelectric film surfaces; critical exponent characterizing P versus E response at T_C
D	Displacement vector; dimension of domain (1, 2, or 3)
d	Film thickness
d_d	Depletion width
Δ	Trap energy in a ferroelectric, measured down from conduction band
$\Delta\Phi$	Image field reduction term
d'	Distance between dendritic microshort tips extending from cathode and anode
dc	Direct current
DRAM	Dynamic Random Access Memory

e	Charge of the electron
E	Electric field
ε	Relative dielectric constant
E_{B}	Breakdown field
E_{c}	Coercive field
EBDW	Electron Beam Direct Writing
ECR	Electron Cyclotron Resonance
EEPROM	Electrically Erasable Programmable Read-Only Memory
EXAFS	Extended X-ray Absorption Fine Structure
F	Helmholtz free energy
FeRAM, FRAM	Ferroelectric Random Access Memory
FET	Field Effect Transistor
Φ	Metal work function
Φ_{B}	Schottky barrier height
Φ_{FE}	Ferroelectric work function
γ	Critical exponent characterizing isothermal susceptibility slightly above T_0
γ'	Critical exponent characterizing isothermal susceptibility slightly below T_0
h	Planck's constant
i	Displacement current density
J	Leakage curent density
j	Exponent characterizing field dependence of switching time
JFET	Junction FET
k	Exponent characterizing frequency dependence of coercive field
K	Thermal conductivity
k_{B}	Boltzmann's constant
K_{s}	A ratio of surface charge density to remanent polarization useful for parametrizing the read operation of a ferroelectric memory
λ	Electron mean free path
L	Ratio of maximum displacement current $i(t)$ times the time at which it is maximum t_{max}, divided by spontaneous polarization P_{s}
m	Exponent characterizing temperature dependence of switching
m^*	Effective mass
MBE	Molecular Beam Epitaxy
MOD	Metal-Organic Decomposition
MOSFET	Metal Oxide Silicon FET
μ	Mobility
n	Carrier concentration

N	Number of microshorts or nucleation sites or space-charge density per unit area
N_s	Number of bipolar switching cycles
NDRO	NonDestructive ReadOut
NV	NonVolatile
P_+, P_-	Delta-function polarization at ferroelectric surfaces
P_i	Polarization due to charged defects
P_r	Remanent polarization
P_s	Spontaneous polarization
p_{tr}	Trapped hole density
PLD	Pulsed Laser Deposition
PLZT	PZT with lanthanum doping
PT	Lead titanate $PbTiO_3$
PZT	Lead zirconate titanate $PbZr_xTi_{1-x}O_3$
Π	Pressure
Q	Quality factor
q	Total electric charge
R	Nucleation rate (sites per cm^3 per second)
RAM	Random Access Memory
RBS	Rutherford Back-Scattering
READ	To decode and output the information stored in a memory
RIE	Reactive Ion Etching
RTA	Rapid Thermal Annealing
ρ	Charge per unit volume
σ	Electrical conductivity
σ_0	Electrical conduction in the high-temperature limit
SBN	Strontium Bismuth Niobate $SrBi_2Nb_2O_9$
SBT	Strontium Bismuth Tantalate $SrBi_2Ta_2O_9$
SBTN	Strontium Bismuth Tantalate-Niobate $SrBi_2Ta_{2-x}Nb_xO_9$
SCLC	Space Charge Limited Currents
SRAM	Static Random Access Memory
τ	Reduced temperature $(T - T_C)/T_C$
$\tan\delta$	Loss tangent
t_B	Breakdown time
t_c	Rise time for increasing film temperature dT/dt or applied field dE/dt
T_C	Curie temperature
TDDB	Time Dependent Dielectric Breakdown
TEM	Transmission Electron Microscope
t_N	Nucleation time
t_s	Switching time
U	Potential energy
UPS	Ultraviolet Photoelectron Spectroscopy

v	Domain wall speed
V	Voltage
WRITE	To encode and input data into a memory
XPS	X-ray Photoelectron Spectroscopy
XRD	X-Ray Diffraction
ξ	Correlation length in bulk for ferroelectric polarization
YBCO	Yttrium Barium Copper Oxide
z	Depth into film from surface or interface

1. Introduction

Since ferroelectricity was discovered in 1921 it has been obvious to many scientists and engineers that the two stable polarization states $+P$ and $-P$ could be used to encode the 1 and 0 of the Boolean algebra that forms the basis of memory and logic circuitry in all modern computers Yet until very recently, this has been unsuccessful. In fact, although ferroelectric materials are used in a wide variety of commercial devices, it has until now always been the case that some other property of the material – especially pyroelectricity or piezoelectricity – is the characteristic actually employed. Ironically, no devices using ferroelectrics have actually required ferroelectricity to work.

Within the past five years this has abruptly changed. The impetus has come from two directions. For Dynamic Random Access Memories (DRAMs) the ferroelectric materials are used in thin-film form as capacitors. Ferroelectrics usually have large dielectric constants ($\varepsilon = 100$ to 1000), in comparison with numbers of order 5 for silicon dioxide (employed in most DRAMs because it is readily deposited in controlled thicknesses simply by free oxidation of the Si surface) or 25 for Ta_2O_5. The density of present day DRAMs is limited primarily by capacitor area per cell (bit). The resistors and transistors take up relatively little area, but the capacitors limit chip bit-density. If the dielectric constant of a capacitor is 100 times larger, then its area can be 100 times smaller, so that the present limits requiring complicated stacking and trenching to maximize surface area can be relaxed.

For Non-Volatile Random Access Memories – NV-RAMs (in which the stored information is retained even if power to the chip is interrupted), ferroelectrics serve not just as capacitors, but as the memory element itself. Their principal advantages in this application are: low-voltage (1.0 V) operation; small size (about 20% of a conventional EEPROM cell) – and cost is proportional to size once high-yield production is achieved; radiation hardness (not just for military applications, but for satellite communications systems); and very high speed (60 ns access time in commercial devices, sub-nanosecond in laboratory tests on single cells). In comparing speeds for computer systems applications it is good to keep in mind that the interconnect time of about 200 ps required for the output of one cell to charge up an adjacent cell is often the rate-limiting parameter, not the switching time of an individual cell. In a typical 200 ns ferroelectric RAM the memory spends 60 ns selecting the addressed cell and reading the data, 20 ns writing the reference 0, 30 ns adding an offset voltage, and about 100 ns getting the data out. The access time is therefore not greatly improved by improving the fundamental switching time of ferroelectric domains.

In the present book I try to combine the present knowledge of device physics for ferroelectric thin-film memories with information about circuit design, materials deposition and characterization, and testing protocols. It is my hope that this will serve as a useful reference for engineers and device physicists in industry and in government laboratories, and as a book that will assist the teaching of this topic in senior or graduate-level engineering and applied physics university classes.

1.1 Basic Properties of Ferroelectrics: Bulk Materials

This is not a textbook on ferroelectricity. However, a review of the basic concepts [1, 2, 3, 4, 5] may be helpful for some readers; and in addition it gives the author the opportunity to correct some serious errors that are found in most texts.

Ferroelectrics are pyroelectric crystals (including fine-grained ceramics) whose spontaneous electric polarization can be reversed by application of an external electric field (coercive field E_c) that is smaller than the breakdown field E_B. Pyroelectricity is a symmetry property of crystals; but ferroelectricity is a practical engineering property, since some pyroelectrics will break down electrically (short) before the coercive field is reached, and this is often a matter of crystal purity and conductivity.

Ferroelectrics are in thermal equilibrium in each polarization state, $+P$ and $-P$. Amorphous materials, including electrets, cannot be ferroelectric. Claims to the contrary either result from different definitions of ferroelectricity than that used here [5] or have confused [6] electret effects [7] or hysteretic properties of the electrode interfaces with true ferroelectricity.

It is necessary but not sufficient that ferroelectrics display an electric hysteresis in their $D(E)$ (or, equivalently, $P(E)$) relationship, as shown in Fig. 1.1a, where D is the electric displacement vector, E the applied electric field, and P the polarization.

Ferroelectrics are generally much more complicated than ferromagnets. This occurs for two reasons. First, most magnets have spatially localized spins whose magnitude is fixed; the difficulties in calculating mean values of observable quantities has to do with the mathematics of averaging these spin variables at different temperatures T and applied fields B. By comparison, the dipoles that give rise to an electrical polarization are not spatially localized; they extend over distances of order 1 nm, and they change with temperature. Second, measuring the electric properties of a ferroelectric involves putting contacts on it and making a circuit. All conventional measurements are therefore those of a *system*, with electrodes and interfaces, and rather

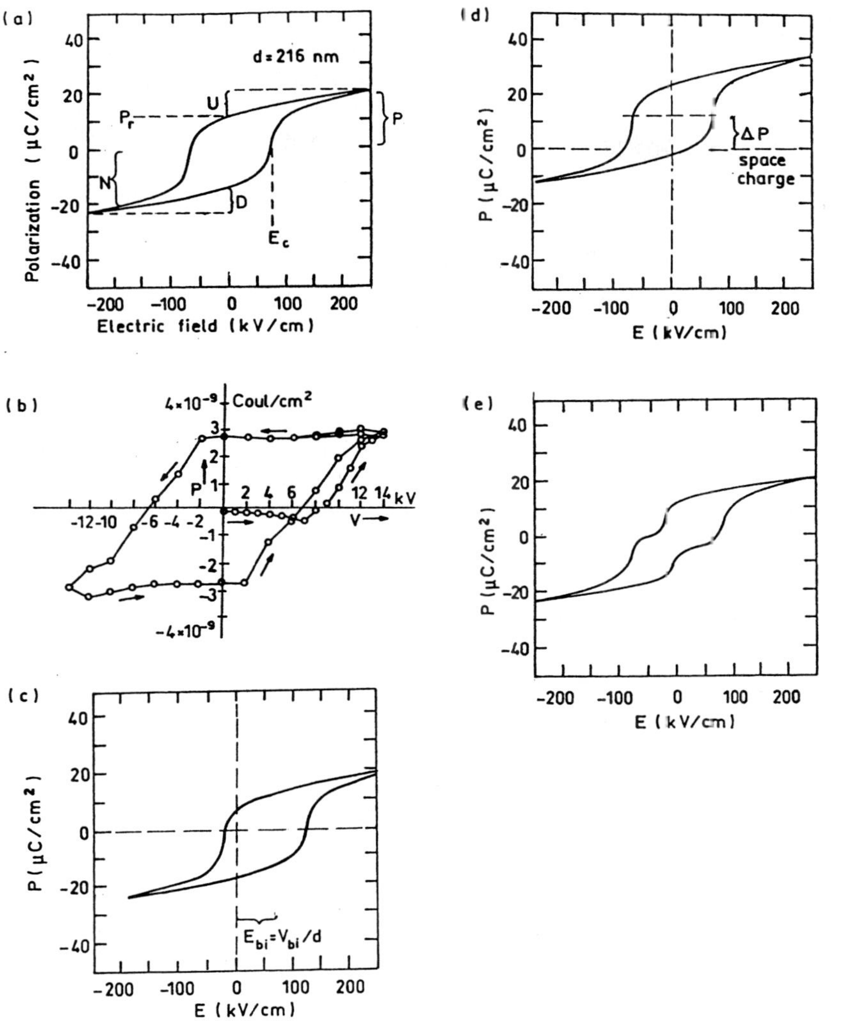

Fig. 1.1. $P(E)$ hysteresis. (**a**) For a ferroelectric (an actual PZT/Pt sol–gel film trace from Panasonic); here the symbols P, U, N, D stand for, respectively: P = Positive switching voltage; U = 'up' non-switching voltage; N = negative switching voltage; and D = 'down' non-switching voltage. (**b**) For an electret. (**c**) For a ferroelectric with a built in bias field V_{bi}, perhaps due to different top and bottom electrodes (horizontal shift). (**d**) For a ferroelectric with trapped space charge at an electrode–ferroelectric interface (vertical shift). (**e**) For a ferroelectric with a damaged surface layer (steps in hysteresis curves)

complicated interfacial physics. In contrast, large magnetic fields can be applied to ferromagnets in vacuum with no physical contacts. The contacts are very important with regard to the hysteresis behaviour. If the top and bottom electrodes are of different materials, there will be an asymmetry in the hysteresis loop; usually this is along the horizontal voltage axis. A particularly strong example is in the Ag/$BaTiO_3$/Pt study of Su et al. [8]. These asymmetries are external and not very interesting, but similar horizontal shifts can arise from built-in bias voltages intrinsic to the material. And vertical shifts along the P-axis can arise from space charge at the ferroelectric–electrode interface. Thus it is generally unwise to use different top and bottom electrodes on samples for which quantitative studies of hysteresis curves are intended, although these may be essential for ferroelectric FETs (Chap. 12).

Hysteresis in a crystal's $P(E)$ relationship is necessary but not sufficient to establish ferroelectricity. A good counterexample is the hysteresis shown in Fig. 1.1b for carnauba wax [7]. Many plastics and waxes are electrets. Electrets are not ferroelectric and need not be crystals; they are materials with large densities of mobile charged defects in a viscous environment. These charged defects can be *poled*; that is, they can be moved to either the top or bottom surface by application of a large electric field, and this poling is reversible, so that an electrical hysteresis results. However, these poled states are not in thermal equilibrium and will relax back to an equilibrium ground state with time (Fig. 1.2). This relaxation can be from a net polarization P back to zero, which is referred to as a *homocharging* effect; or, as shown in Fig. 1.2, they can actually reverse polarization via relaxation, an effect known as *heterocharging*. Therefore, unlike ferroelectrics, they will not be stable enough to be usable as computer memories. Moreover, their hysteresis properties are not intrinsic (unlike those in ferroelectrics) but arise entirely from defect chemistry.

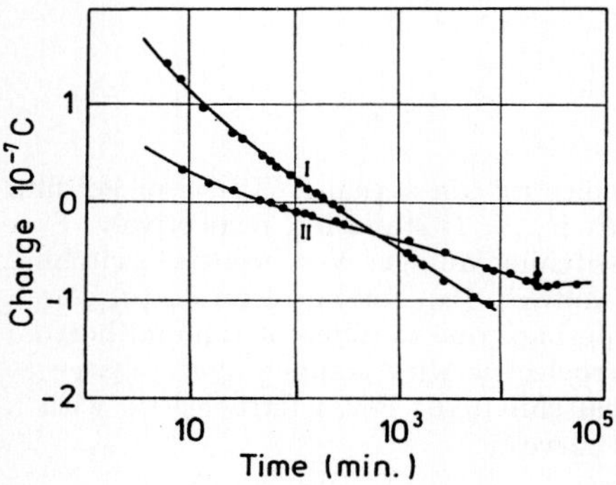

Fig. 1.2. Electret relaxation. Polarization versus time (two different specimens) [7]

The basic properties of ferroelectrics arise from the fact that at least one set of ions in the crystal sits in double-well potentials (Figs. 1.3, 1.4). At low temperatures all such ions in a local region of the crystal sit on the same side of the double well. This region is called a domain. In a ceramic it may be larger than or smaller than a grain. That is, domain walls do not coincide with grain boundaries. The fact that all ions in a domain sit on the same side of their double wells gives rise to a net electric dipole per unit volume, called a polarization P. In a material that is not specially prepared, there will be many domains of random polarization, and the macroscopic polarization mean value $\langle P \rangle$ will be zero. If, however, the sample is subjected to a large electric field E, the ferroelectric will be *poled*; that is the dipoles will preferentially line up along the applied field, and there will be many more $+P$ domains than $-P$ or with P at angles to the field E. It is not true that charged defects will become poled at the same applied field as the domains do. Therefore, there can be situations in which the charged defects have a memory of an initial high-field polarization, while the domains switch back and forth at subsequent bipolar voltage cycles. This memory effect is usually called 'imprint' [9] implying a preferential polarization state for domains, usually the one into which they (and their surrounding defects) were first poled. This term is intentionally analogous to the 'imprint' or preference a duckling has for the first creature it sees upon hatching (initial orientation).

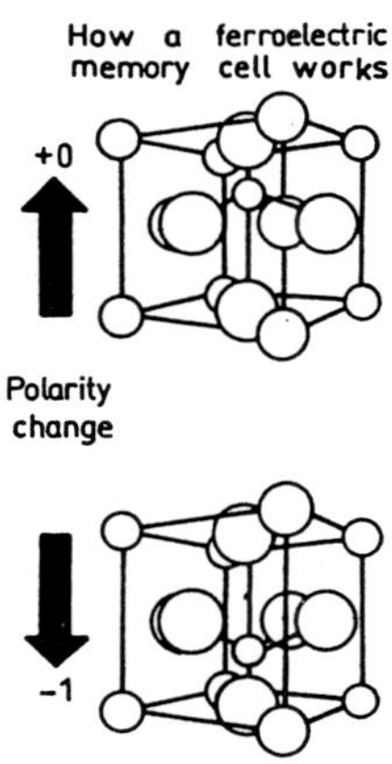

Fig. 1.3. Displacements in ferroelectric $BaTiO_3$

As temperature is increased, the particles in the double wells (usually hydrogen ions, or oxygen, or fluorine) obtain enough kinetic energy to hop rapidly back and forth between the two sides of the double well, so that their time averaged position $\langle x \rangle$ is midway between the initial two sites. This can happen for two different reasons. If the shape of the double well $U(x)$ remains independent of T but the ion kinetic energy $k_B T$ becomes large, the phenomenon is called an *order–disorder* phase transition. The ion positions are ordered below a phase transition temperature T_0 and disordered above. However, if the shape of the potential $U(x, T)$ changes rapidly with temperature T, its minima may actually move together to their initial midpoint. This

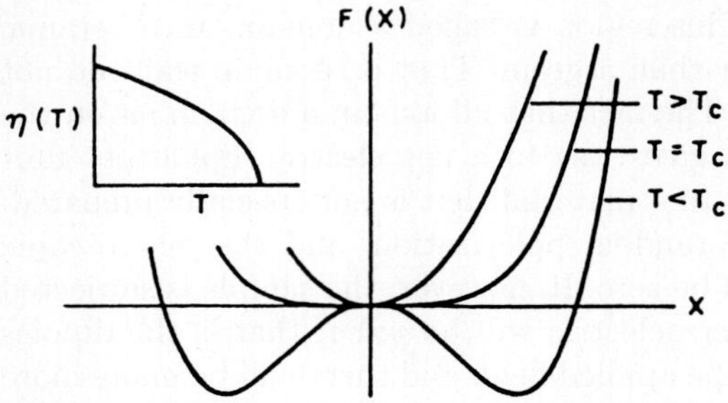

Fig. 1.4. Potential $U(x, T)$ at different temperatures for a second-order transition ($B > 0$). *Inset*: expectation value of polarization $\langle P \rangle$ (which is proportional to expectation value $\langle x \rangle$) versus temperature

can happen over a small temperature range for which the ion kinetic energy $k_B T$ changes very little. This kind of phase transition is called *displacive* and is the subject of Landau's 1937 general theory [10] and the 1949 theory of Devonshire, who independently developed it explicitly for ferroelectrics [11]. It was once thought (and taught in textbooks) that hydrogen-bonded ferroelectrics such as KH_2PO_4 (KDP) or Rochelle salt were order–disorder and that other ferroelectrics (e.g., oxides such as $BaTiO_3$) were more displacive. This was driven by an important 'red herring' (false clue): it was well known that fully deuterating KD_2PO_4 raised its transition temperature by around 100 K, an effect interpreted theoretically as evidence for proton (or deuteron) tunneling between the two wells of the double-well potential on the O-H–O hydrogen bond [12]. In fact, this example was taken for several decades as the only example of tunneling of a particle heavier than an electron in solid state physics. KDP was a true paradigm [13]. Unfortunately, recent studies [14] have shown that the difference in transition temperatures between KH_2PO_4 and KD_2PO_4 arises primarily from the difference in zero-point energies for O-H and O-D bond motion and the fact that the O-H–O bonds are not colinear, but slightly bent. The phenomena near T_C are not primarily those of proton tunneling, then, but of a displacive phase transition[1]. (Some clear evidence for proton tunneling is manifest, however, at low temperatures by the T-independence of the spin-lattice relaxation time T_1 [R. Blinc, private communication].) Other spectroscopic evidence [15] shows strongly that these ferroelectrics are displacive, and not fundamentally order–disorder tunneling systems. The same is true in Rochelle salt and in [16] tris-sarcosine calcium chloride (TSCC), where earlier NMR data had been misinterpreted [17]. We

[1] This question of tunneling might have been resolved thirty years ago. In 1971 the present author initiated experiments in W-division at Los Alamos (with the enthusiastic support of Berndt Matthias) to grow fully tritiated KT_2PO_4 powder from T_2O heavy water and to measure its Curie temperature via specific heat. These experiments were never completed due to safety concerns. Although safety requirements are now much stricter, this experiment should be redone.

conclude that most ferroelectrics are quite displacive and thus can be well described by Landau–Devonshire theory and that none are dominated by any kind of quantum mechanical tunneling between double-well minima in a classic order–disorder system. What appears to happen instead is that subtle displacive rearrangements of ions greatly influence the degree of ordering of other ions.

1.1.1 Landau–Devonshire Theory

In the basic Landau–Devonshire theory one assumes that the Helmholtz free energy F can be expanded in a power series in the order parameter for the system. For a ferroelectric, the macroscopic order parameter is the polarization $P(T)$; the microscopic polarization is the displacement x of a set of ions from their position midway along some double well potential, for example, the Ti ions in $BaTiO_3$ (Fig. 1.3). It is assumed that $\langle P \rangle$ is proportional to $\langle x \rangle$. The free energy is assumed to have the explicit form

$$F(P, T, E) = -EP + c + AP^2 + BP^4 + CP^6 , \tag{1.1}$$

where c is a constant of no importance and the coefficient A has the form

$$A = A_0(T - T_C) , \tag{1.2}$$

where T_C is called the Curie temperature and is not necessarily equal to T_0, the phase transition temperature.

The assumed form of (1.2) is a necessary result of mean field theory. In this theory it is assumed that every ion in a double well in the crystal interacts with every other such ion equally, independent of how far away it is. This is of course a very crude, rather unphysical assumption, but it permits us to replace all the ions except one with an average or mean field acting on the first ion. This makes complex mathematics exactly solvable. Often this mean field assumption is buried in theories and not obvious to the student. The van der Waals theory of liquid–gas transitions is mean field. The Weiss theory of magnetism is mean field. Ironically, gases and magnets are experimentally not mean field systems at all, because the interactions among molecules or spins are very short range. But two systems in nature are very mean field indeed: superconductors and ferroelectrics. In each case the interactions are very long range. In ferroelectrics the interactions are largely Coulombic, dropping off slowly as $1/r$. In superconductors the Cooper pairs are not physically close together (no matter how they are drawn in textbooks!).

An important aspect of mean field theories is that certain relationships among thermodynamic quantities near T_0 are the same in all such systems, independent of microscopic details; thus they are the same in the Weiss theory of magnetism, the Van der Waals theory of liquids, and the Devonshire theory of ferroelectrics.

The free energy of any system is minimized in thermal equilibrium. Minimizing F in (1.1) with respect to P yields $P(T)$. Two cases are of interest: if the coefficient B is positive we can neglect the coefficient C in (1.1) and find

$$P(T) = P(0)\left(\frac{T_C - T}{T_C}\right)^{1/2} , \tag{1.3}$$

in which case the phase transition is continuous (second order) and $T_C = T_0$. In this case there is no latent heat of transition.

If $B < 0$ the phase transition is discontinuous (first order). The evolution of the potential $U(x,T)$ is shown for both cases in Figs. 1.4 and 1.5, and the evolution of $P(T)$ and $x(T)$ is shown in both cases in the inserts in these two figures. If $B < 0$ then T_C is not equal to the transition temperature but is higher, if extrapolated from below. This is most readily seen by examining the behaviour of the dielectric constant (permittivity) versus temperature, as shown in Fig. 1.7.

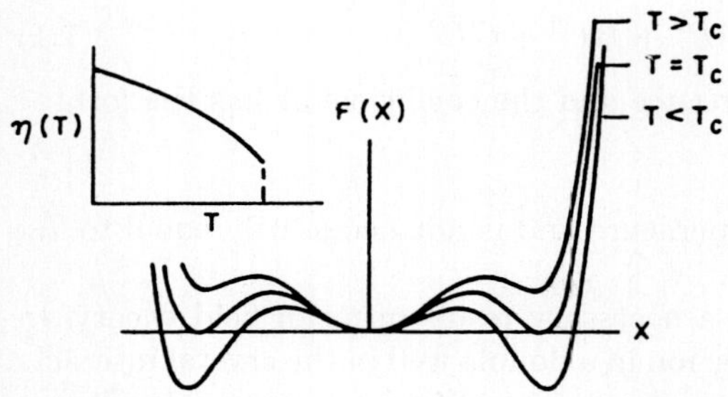

Fig. 1.5. Potential $U(x,T)$ at different temperatures for a first-order transition ($B < 0$). *Inset*: expectation value of polarization $\langle P \rangle$ (which is proportional to expectation value $\langle x \rangle$) versus temperature

The dielectric constant is denoted as $\chi(T)$ or $\varepsilon(T)$ and defined by

$$P(E,T) = \chi(T)E , \tag{1.4}$$

where E is the applied field. Using (1.3) we see that

$$\chi(T) = \chi(0)\frac{T_C}{T_C - T} \tag{1.5}$$

for a bulk ferroelectric, and hence

$$C^{-1}(T) = \chi^{-1}(T)d/A . \tag{1.6}$$

However, for a thin film, $1/C(T)$ will not be given by (1.6), but by two capacitances in series, the bulk contribution from the interior of the film, and a surface term C_i from the electrode–ferroelectric interface:

$$C^{-1}(T) = C_i^{-1} + \chi^{-1}(T)d/A . \tag{1.7}$$

Usually the interfacial capacitance is not temperature dependent and often arises from a damaged or amorphous layer.

The product $\chi(0)T_C$ is often called the Curie constant and is usually designated C [not to be confused with capacitance C or the sixth-order coefficient in (1.1)]. The fact that the electric susceptibility diverges as $(T_C - T)^{-1}$ both above and below the phase transition temperature (Fig. 1.6) is an important characteristic of a mean field phase transition. Keeping in mind that T_C is not equal to the transition temperature T_0 for a first-order transition, we must define a value T_C for susceptibilities fitted to data at $T > T_0$, and a different value T'_C for data fitted to (1.8) below T_0:

$$\chi(T) = \chi(0)\frac{T'_C}{T - T'_C} \,. \tag{1.8}$$

Typically T_C and T'_C may differ from T_0 by 10 K to 100 K. Equations (1.5–1.8) are termed Curie–Weiss Laws after the analogous effects in magnetism. In more general terms, one defines for all phase transition systems

$$\chi(T) = \chi(0)\left(\frac{T_C}{T - T_C}\right)^{\gamma} \,. \tag{1.9}$$

In mean field $\gamma = 1$ exactly. For $D = 2$ Ising models, $\gamma \approx 5/4$. For $D = 3$ Heisenberg models, $\gamma \approx 1.4$. Our work shows that $\gamma = 1$ for ferroelectrics [18, 19, 20] and I believe that there are few if any exceptions among the ferroelectrics of interest to the memory-device community.

Some textbooks claim that there is a simple 2:1 ratio of the slopes of $\chi^{-1}(T)$ versus T above and below T_0 in mean field theory, and that getting a different result (say 5:1, as in LaP_5O_{14}) shows (Fig. 1.7) that the system is not mean field. This is *not* true. It happens only for the simplest possible free energy expression (1.1) but not if strain coupling terms are included. This is discussed correctly in great detail by Errandonea [21]. Failure to get a 2:1 ratio for the slopes of inverse dielectric constant versus temperature above and below the phase transition temperature can *never* be used to imply failure of mean field theory in (1.2). It is (1.1) that is too simple, not (1.2).

Finally, we might ask what the transition temperatures in ferroelectrics should be – hundreds of degrees, degrees, or millidegrees? Figure 1.8 shows [21] that for 339 well studied ferroelectrics the transition temperatures follow roughly a log-normal distribution, with a maximum near room temperature. This may understood qualitatively be recognizing that the energy required to remove an ion completely from a double well potential (breaking a bond) is of order 1 eV, so that the height of the potential barrier within the well in Fig. 1.3 must be an order of magnitude smaller. 0.1 eV is about 800 K, so very roughly, T_C should be most commonly in this range of temperature.

1.1.2 Soft-Mode Theory

Beginning in 1960 Cochran showed [24] that the displacements required to transform the high-temperature, more-symmetry paraelectric crystal lattice

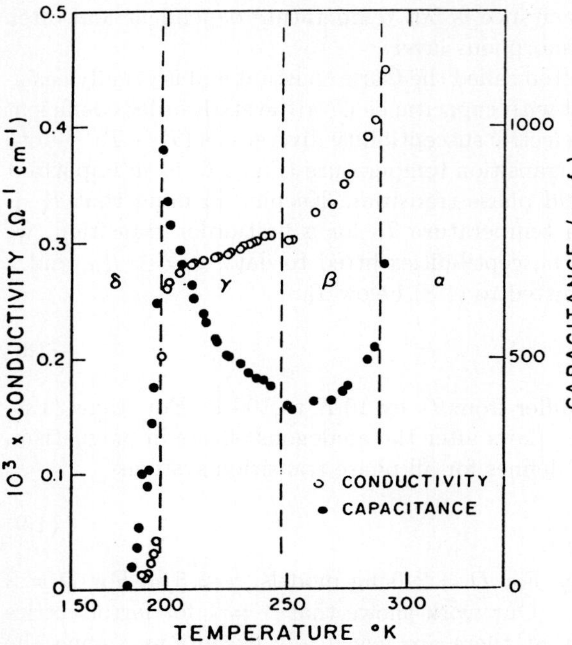

Fig. 1.6. Susceptibility $\chi(T)$ versus temperature near T_0 for an oxide ferroelectric ($Ag_{13}I_9W_2O_8$) [22]

into the lower-symmetry ferroelectric phase correspond to an eigenvector of a single normal mode in cases where the phase transition is second order (continuous). The force constants for this mode literally soften as the phase transition temperature is approached from above or below. Only the transverse optical mode of long wavelength ($q = 0$) softens for a ferroelectric, and this arises because of a subtle cancellation of long-range Coulombic forces and short-range forces R:

$$\mu\omega^2 \,(\mathrm{TO}) = R - \frac{4\pi(\varepsilon + 2)(Ze)^2}{3V(T)} \,, \tag{1.10}$$

where μ is a vibrational normal-mode reduced mass, ε is the optical dielectric constant ε_∞, Z is the effective charge for the mode, and V is the cell volume, which increases with T due to anharmonic thermal expansion. Small changes in V with T cause a cancellation of the two terms on the right hand side of (1.10) at

$$V(T) = \frac{4\pi(\varepsilon + 2)(Ze)^2}{9R} \,. \tag{1.11}$$

By comparison, the longitudinal optic mode frequency consists of two positive terms and cannot vanish:

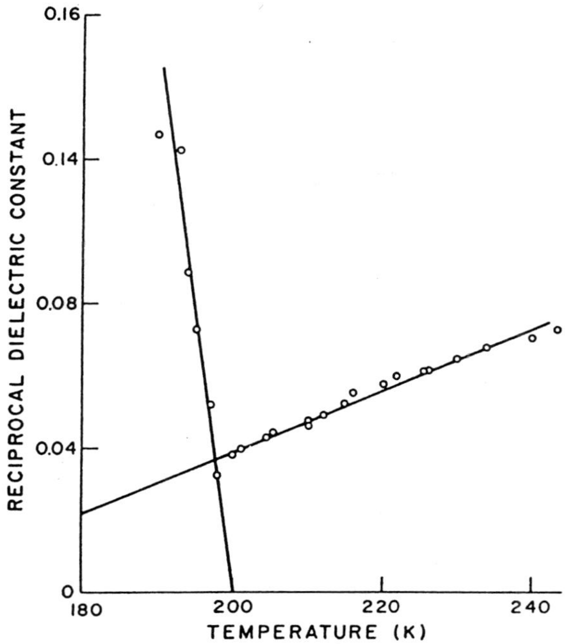

Fig. 1.7. Inverse susceptibility $\chi^{-1}(T)$ near T_0 for $Ag_{13}I_9W_2O_8$. The ratio is about 5:1 below/above T_0 [23]

$$\omega^2\,(\mathrm{LO}) = R + \frac{8\pi(\varepsilon+2)(Ze)^2}{9eV(T)} \,. \tag{1.12}$$

This model describes many ferroelectric phase transitions, as reviewed by the present author [24] in 1974. It is a displacive model and it does not, however, describe purely order–disorder mechanisms, such as the phase transition in NH_4Cl. Other models of ferroelectricity have also been proposed that do not involve soft optical phonons: the first model in which polarization was generated largely from mixing of electronic wave functions was Jaynes' PhD thesis under Wigner at Princeton in 1950 [24]. With this model they could fit both the Curie constant C above T_C and the polarization P_r below T_C with a single parameter. This parameter is not freely adjustable, because it corresponds to the energy of the lowest electronic state above the ground state. Jaynes calculated this in $BaTiO_3$ as 0.13 eV. Unfortunately, no such intrinsic electronic level is found in barium titanate, and so the 'electronic' model fails. This model has been resurrected in a long series of papers by Kristoffel and Konsin, with a notable lack of success in quantitatively explaining real ferroelectric phase transitions.

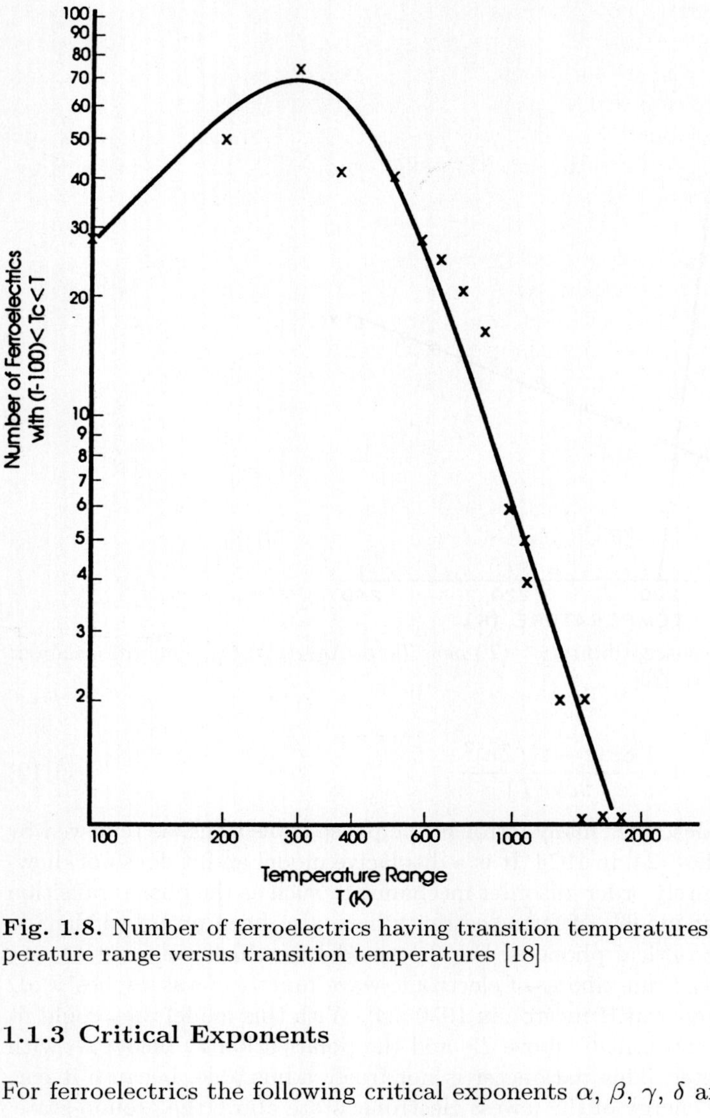

Fig. 1.8. Number of ferroelectrics having transition temperatures in a 100 K temperature range versus transition temperatures [18]

1.1.3 Critical Exponents

For ferroelectrics the following critical exponents α, β, γ, δ are defined:

$$
\begin{aligned}
C_\mathrm{V}(T) &= \text{constant } (T_\mathrm{C} - T)^{-\alpha} \, , \\
P_\mathrm{r}(T) &= \text{constant } (T_\mathrm{C} - T)^{\beta} \, , \\
\chi(T) &= \text{constant } (T_\mathrm{C} - T)^{-\gamma} \, , \\
D(E) \ \text{ at } T &= T_\mathrm{C} \cong P(E) = E^{1/\delta} \, .
\end{aligned}
\tag{1.13}
$$

For second-order transitions $T_\mathrm{C} = T_0$ and the numerical values for α, β, γ, δ in mean field are $\alpha = 0$ (logarithmic divergence), $\beta = 1/2$, $\gamma = 1$, and $\delta = 3$.

Primes on α and γ denote values below T_C; unprimed, above T_C. Very recently Perez-Mato (EMF-9, Prague, July 1999) has shown that β measured in ferroelectrics and in antiferroics (e.g., $SrTiO_3$) is not a universal critical exponent. Values of $\beta = 0.3$ may be obtained for the temperature dependence of the order parameter in such cases but be perfectly mean field. This surprising result arises from the fact that such structural phase transitions are probably not in the displacive limit; even a small amount of order–disorder character results in a mean field exponent that can lie anywhere from zero to 0.5, with 0.3 being quite plausible and perhaps even common. Thus, readers should be advised to take all values of β with a 'grain of salt': they do not necessarily imply true critical (fluctuation-dominated) behaviour. Similarly, γ can be greater than 1.0 without critical effects, as measured in LaP_5O_{14} by Errandonea [21] and by Chen and Scott [19, 20].

1.1.4 Tricritical Points

Under the special circumstances in which the quartic coefficient B in (1.1) is exactly zero (for example, because of pressure), the phase transition becomes *tricritical.* That is, there is a point in the three-dimensional phase diagram of temperature, pressure, and applied field (Fig. 1.9) at which the phase transition goes from second order (continuous) to first order (discontinuous). This is called a *tricritical* point because there are three lines of continuous phase transition boundaries that come together at a single point.

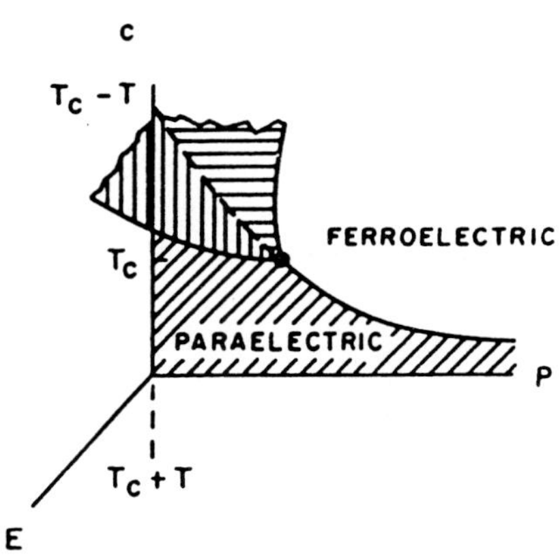

Fig. 1.9. Tricritical point phase diagram

Many common ferroelectrics are very near tricritical points under ambient conditions. In mean field theory at tricritical points most critical exponents differ from those at ordinary second-order phase transition points: $\alpha = \alpha' = 0$, $\beta = 1/4$, $\gamma = \gamma' = 1$, and $\delta = 5$.

Landau theory fails at tricitical points in one respect: it is not compatible with the Griffiths inequalities [25]

$$\alpha' + \beta(1+\delta) \geq 2\,, \tag{1.14}$$

$$\gamma(\delta+1) \geq (2-\alpha)(\delta-1)\,, \tag{1.15}$$

nor with the Rushbrooke inequality [26]

$$\alpha' + 2\beta + \gamma' \geq 2\,. \tag{1.16}$$

Since these inequalities derive from thermodynamics they transcend the specific details of Landau–Devonshire theory. Note that they are all satisfied for ordinary second-order transitions ($\alpha = 0$, $\beta = 1/2$, $\gamma = 1$, $\delta = 3$). The probable reason for this failure is that α actually equals 1/2 at tricritical points. We note that in scaling theory α' is not an independent parameter but is given by

$$\alpha' = 2 - \beta(\delta+1)\,, \tag{1.17}$$

which is exactly 1/2 at a mean field tricritical point. This is discussed elsewhere [27] and experimental values in some ferroelectrics are indeed $\alpha = 0.4$ or so [27, 28, 29, 30].

It is my opinion that all ferroelectric materials used at present for memory applications are mean field, and that many of them are near tricritical points. Hence, one can expect them to be described by the approximate values $\alpha = 1/2$, $\beta = 1/4$, $\gamma = 1$ and $\delta = 5$ in (1.13–1.16).

In real crystals B is not independent of T or pressure Π. For example, in $BaTiO_3$, B varies linearly with T near T_0, causing (1.1–1.3) to be poor approximations. B often varies strongly with pressure, so that the transition can change from first-order to second-order at a critical pressure Π_0. And for ferroelectric memory engineers it is most important that A, B, and T_C all vary with uniaxial stress in thin films, which can arise from lattice constant mismatch at the dielectric–electrode interface.

Although ferroelectric memories should not be operated near their paraelectric phase transition temperatures, the ideas above may be useful to memory engineers in understanding the overall relationships between P, E, and T and that all common oxide and fluoride ferroelectric films should display mean field behaviour, albeit that near tricritical points in many cases.

The application of this kind of Landau–Devonshire theory to the device physics of ferroelectric thin-film memories is of the utmost importance in incorporating uniaxial stress into the problem. Stress may couple to the polarization in a variety of algebraic ways, thus modifying (1.1). Comprehensive books have been devoted to this complication [31]. Most significant is that the lattice mismatch between substrate and ferroelectric film (or bottom electrode and ferroelectric film) can shift the Curie temperature or even change the entire crystallographic phase diagram, creating new phases that are not present in bulk. This aspect of thin film crystallography has been developed nicely over the past two years by Pompe, Speck et al. [32], with results that are of particular import for PZT thin films. The Landau model for $BaTiO_3$ films epitaxially grown onto cubic substrates, giving distinct phases (so-called c-, a-, and ac-phases) due to misfit strain was developed by Pertsev et al. this year [32].

These authors consider the possible domain patterns in epitaxial pseudocubic ferroelectric thin films. Their basic idea is that the domain structures fit together by shearing single domains. This shear accompanies the paraelectric–ferroelectric phase transition in PZT, which requires an elongation of 2.6% along the *c*-axis (pseudocubic [001] direction). Thus, a band of 90-degree domains will form a zig-zag chain and be connected through (011) faces. This in turn requires a modulation of the PZT surface with periodicity equal to twice the domain width. Individual domains form a *tiling* at the surface and are inclined by 54° towards the usual (111) surface of as-deposited PZT and 45° towards the surface normal. Pompe et al. begin by considering ferroelastic effects and in particular the {100} twin boundaries. A puckered lamellar

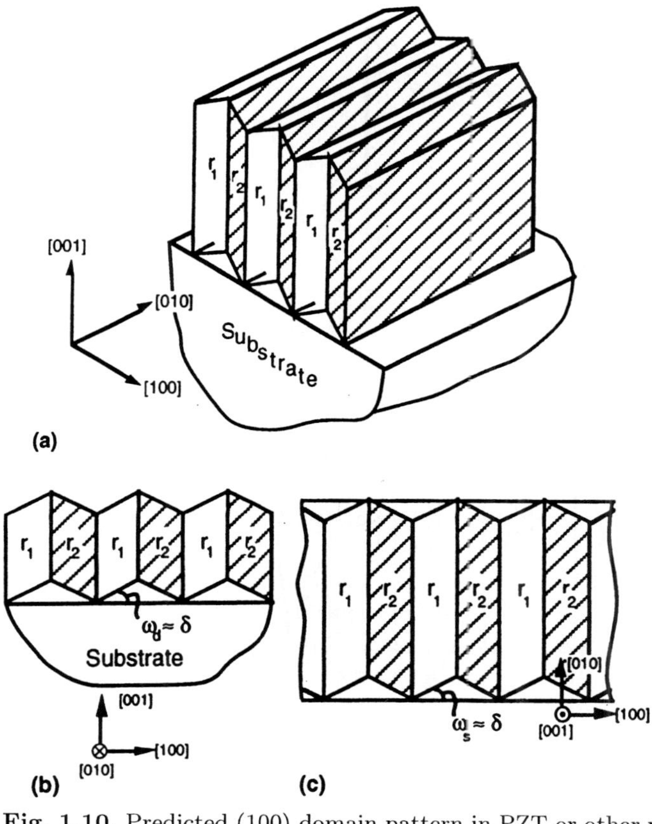

Fig. 1.10. Predicted (100) domain pattern in PZT or other pseudocubic ferroelectric thin films: (a) perspective view, (b) side view, showing puckering, (c) plan view [32]

pattern shown in Fig. 1.10 is predicted. The {110} boundaries are shown in Fig. 1.11. In their second paper they extend this work to include defects. It is of particular importance that these domain patterns in PZT thin films can be imaged by Atomic Force Microscopy techniques (AFM), as first shown by Takashige et al. [32]. Zybill et al. [32] find from their AFM data obtained from Cr/Ni or Ti-metallized PZT films that domain slab widths shown in Fig. 1.10 are 20–27 nm for (100) and (010) planes. For (110) planes (Fig. 1.11) they are even narrower: 10–15 nm. This is remarkably narrow for 90-degree domains, which in bulk are more than an order of magnitude wider. Even the narrower 180-degree domains are not this narrow in bulk. These surface domain details may affect the device behaviour of PZT thin-film memories in many subtle ways.

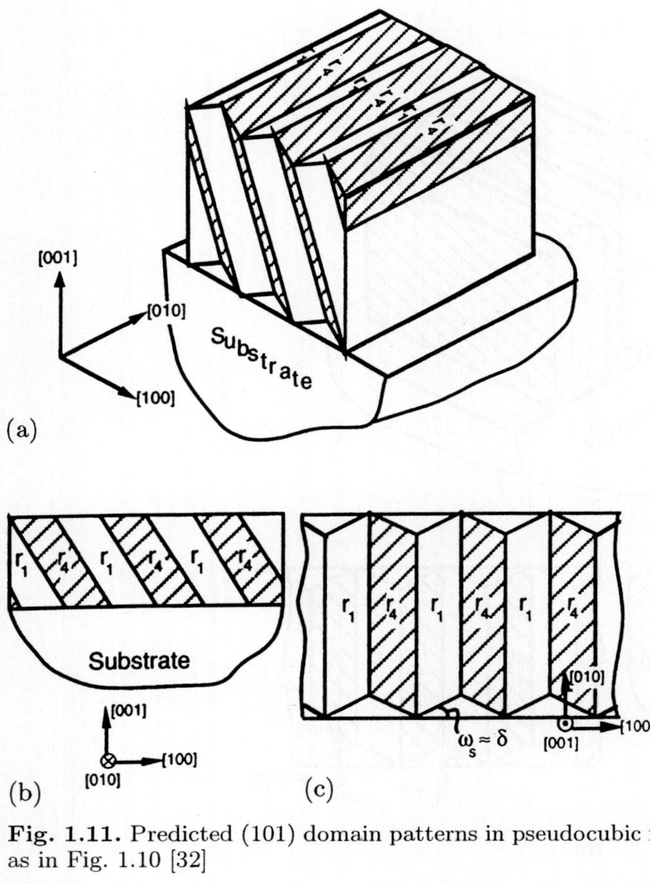

Fig. 1.11. Predicted (101) domain patterns in pseudocubic ferroelectric thin films, as in Fig. 1.10 [32]

1.1.5 Incommensurate Ferroelectrics and ANNNI Models

The Landau–Devonshire theory outlined above is quite satisfactory in materials in which the size of the crystallographic primitive cell remains the same above and below the transition temperature (ferroelectric–paraelectric) or in which the unit cell doubles below T_0 (antiferroelectric–paraelectric). However, many ferroelectrics have more complicated crystallographic phase sequences in which the ferroelectric unit cell at low temperatures is a larger integral (3, 7, ..., 17) times the original paraelectric high-temperature cell, with numerous phases at intermediate temperatures. These intermediate phases may have translational periodicity which is not an integral multiple of the high-temperature paraelectric-phase lattice constant a_0. That is, their translation vector is not *commensurate* with a_0. These systems can in principle be described via extensions of (1.1), but it is neccessary to retain such high-order terms in P (such as P^{17}), that the physical justification of such models seems unrealistic, even silly [33]. However, these systems are well described via Axial Next-Nearest-Neighbor Ising models (ANNNI models), which can predict their sequence of phases upon lowering temperature.

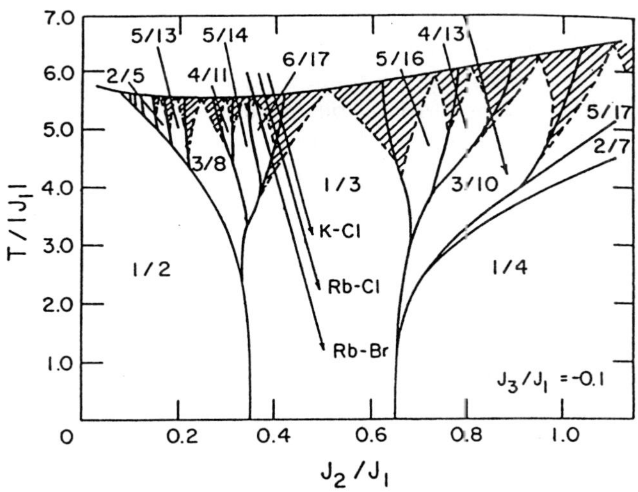

Fig. 1.12. ANNNI model phase diagram for A_2MX_4 ferroelectrics [33]

These ANNNI models represent the interaction of neighboring and next-nearest neighbor unit cells by fictitious exchange constants J and J', in analogy with magnetic models. These models are qualitatively distinct from mean field Landau model equations (1.1) and (1.2). Such a phase diagram is shown [33] in Fig. 1.12 for the K_2SeO_4/Rb_2ZnCl_4 family of A_2BX_4 ferroelectrics. Here temperature is plotted vertically, and the horizontal axis is the ratio J_2/J_1 of next-nearest neighbor exchange constant to nearest-neighbor exchange (both assumed negative). The fractions shown give the reciprocal translational repeat length q_0 as a fraction of the that for the highest temperature phase. Shaded regions are incommensurate where this ratio is irrational (and the translational periodicity is technically infinite). The phase trajectory found experimentally is slightly non-vertical, which allows for the fact that the exchange constant J can be temperature dependent. Unlike the case in magnetism, these exchange constants have no simple physical interpretation and are purely empirical parameters.

These ideas may be of interest for the ferroelectric memory community because Viehland et al. have shown [34] that lead strontium titanate ceramics are incommensurate; and the $BaMF_4$ family used for ferroelectric FETs at Westinghouse [35] also includes incommensurates [36].

1.2 Ferroelectric Films: Depolarization Fields and Finite Size Effects

There has been great interest in the physics of ferroelectric thin films since about 1970. In 1972 Batra and Silverman at IBM published a theoretical calculation [37] that showed that ferroelectric thin films on semiconducting electrodes would be unstable against depolarization fields for film thicklnesses $d < 400$ nm, and on metal electrodes at much thinner values of d. This had the effect of terminating all IBM work on ferroelectric memories [38]. The subject had a minor renaissance a decade later; stimulated by earlier theoretical work on magnetic films [39] and in particular the mean field description of such films [40], Tilley and Zeks wrote a seminal paper [41] analyzing the variation of polarization $P(z)$ as a function of depth z into the ferroelectric and away from the electrode, as well as the change in transition temperature T_0 that results from the non-constant value of $P(z)$; typically the effect is to lower T_0 from that in bulk, roughly by an amount proportional to $1/d$. For a thin ferroelectric film the effects can be large, and theoretically there will be a minimum value of d below which ferroelectricity is not a stable state[2].

The key equation for Helmholtz free energy F per unit area A adds to that in (1.1) a gradient term [42]:

$$\Delta F/A = D_0\xi^2 \left(\frac{\mathrm{d}P}{\mathrm{d}z}\right)^2 , \tag{1.18}$$

where ξ is the polarization correlation length in bulk, plus a surface term

[2] The question of minimum film thickness for ferroelectricity has recently been resolved experimentally (T. Tybell, C. Ahn, and J.M. Trisone, Appl. Phys. Lett. **75**, 856 (1999)) and theoretically (Ph. Ghosez and K.M. Rabe, Appl. Phys. Lett. (2000, in press)). The minimum measured critical thickness is ≤ 4 nm in $PbTiO_3$ and agrees with a total energy calculation that carefully includes image forces in the metal electrodes.

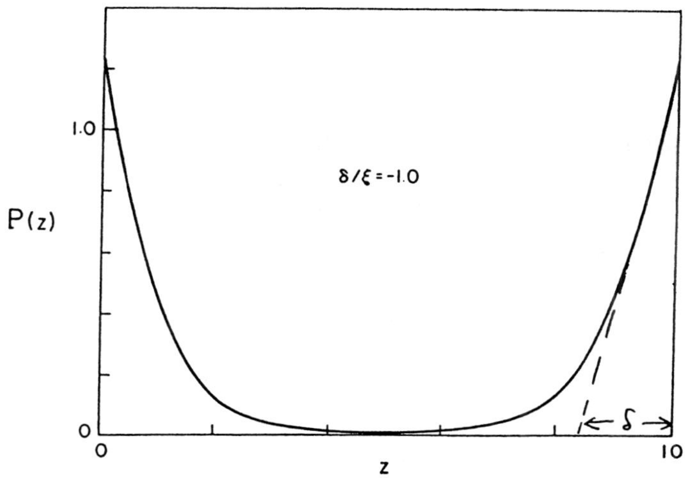

Fig. 1.13. $P(z)$ obtained by integrating by parts (1.1) over dz with the term from (1.18) in it and solving an Euler–Lagrange equation [41]. The *increase* in $P(z)$ near the surfaces agrees with theory and experiment for PZT

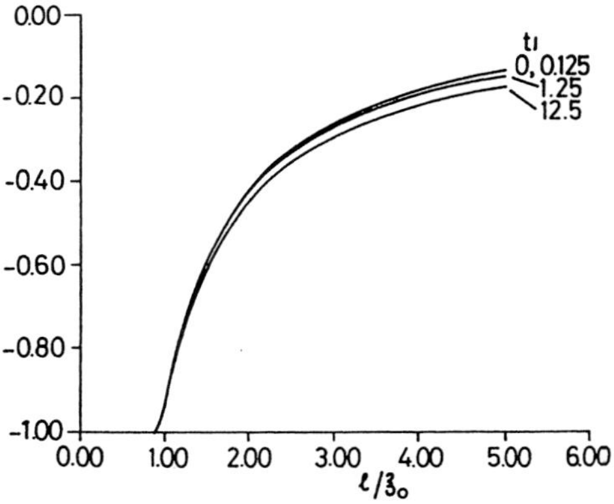

Fig. 1.14. T_0 versus d from $P(z)$ [41]. The ratio of change in T_C over T_C is plotted versus film thickness d, with d in units of the bulk correlation length ξ

$$\Delta F'/A = \frac{\xi^2}{2\delta}\left(P_+^2 + P_-^2\right) , \tag{1.19}$$

where P^+ and P^- are the polarizations at the film surfaces, and δ is an extrapolation length that describes how sharply curved $P(z)$ is near the surfaces. δ is a phenomenological term. Neither its sign nor its magnitude are at present calculable from microscopic models. Figure 1.13 shows $P(z)$ for a small negative δ, measured in units of ξ, and Fig. 1.14 shows how T_0 of the film will vary with d. d is defined in such a way that T_0 and other parameters of interest vary as $1/d$, so maximum effects occur for d small in magnitude (positive or negative). Of particular interest is the fact that for $0 > d > -1$ (units of x), the surface of the film orders at temperatures above that of the

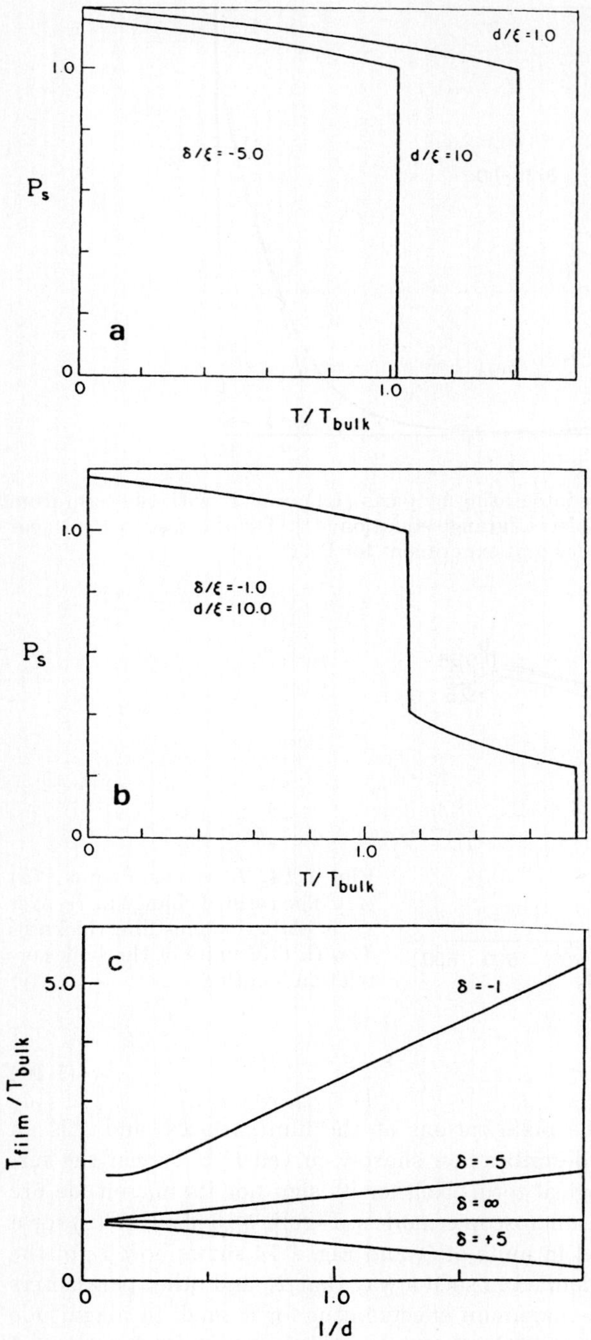

Fig. 1.15. $P(z)$ versus z for $d/\xi = +10$, showing surface ordering: (**a**) $\delta/\xi = -5.0$, showing first order transition at a temperature different from bulk but no surface ordering; (**b**) $\delta/\xi = -1.0$, the threshold for surface ordering above that of the interior of the film; (**c**) film transition temperature versus $1/d$ – note the split-off surface ordering at a threshold of $\delta/\xi = -1.0$ [34]

interior ordering [43], as shown in Fig. 1.15. This is surface ferroelectricity, which may have been observed by Bune et al. [44].

Until very recently it was not possible to extract from experimental data the constants D_0 and δ that appear in the theory of Tilley and Zeks in (1.18). However, this year Chen et al. [45] have made it clear that these parameters can be determined from infrared reflectivity studies quite unambiguously, and an attempt is underway [46] to measure them for SBT at present.

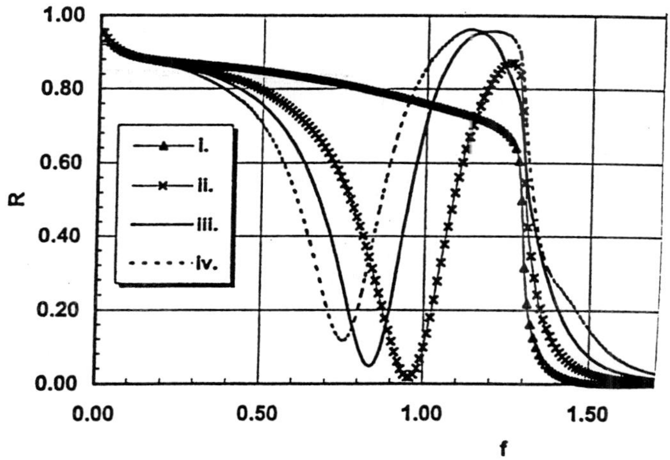

Fig. 1.16. Infrared reflectivity curves calculated [45] for various values of δ. Reflectivity r^2 is plotted versus frequency f. D_0 (and α, see Fig. 1.17) are fixed at $\alpha = 10^{-3}$; damping g is fixed at 10^{-2}. The four curves show $\omega_0\delta/c = 10^3$ (**i**), 10 (**ii**), 1 (**iii**), and 0.1 (**iv**). *Insets* show $P(z)$ at $T = 0.5T_C$ and $0.8T_C$

Figure 1.16 shows the effect [45] in infrared reflectivity curves of progressively decreasing δ (from an initially infinite value corresponding to constant $P(z)$); and Fig. 1.17 shows the effect of varying D_0. In the latter case, the effect is mixed in a way that involves the damping constant, but the two are separable.

Thus it appears that experimental tests and data-fitting to determine the parameters in a detailed theory of $P(z)$ in ferroelectric thin films are now possible.

A rather different approach has been taken by Watanabe [47], who has emphasized the effects of depolarization fields in a model which (correctly) treats the ferroelectric itself as a wide bandgap semiconductor. His results show explicitly that realistic inclusion of surface/interface traps may permit further reduction in minimum film thickness for stable ferroelectricity, but at the cost of very slow switching times. This may explain the extremely slow switching times (seconds) observed [44] in films only a few [nm] thick.

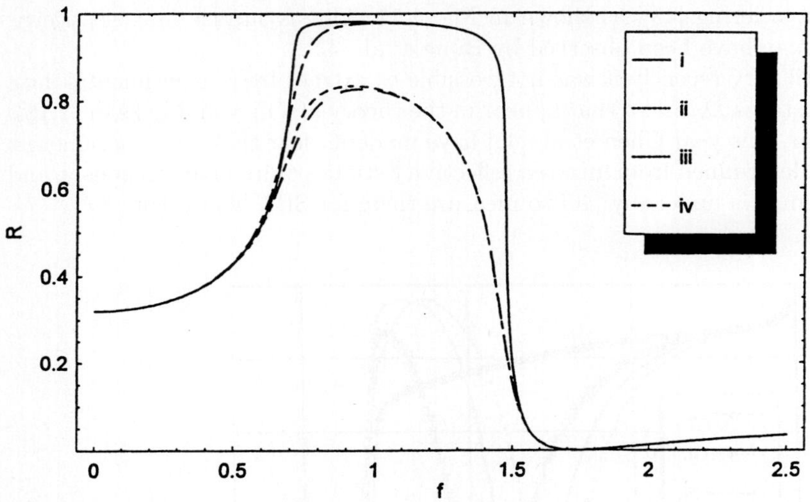

Fig. 1.17. Infrared reflectivity curves calculated [45] for various values of D_0. All curves are at $T = 1.5T_C$. The top two curves (**i** and **iii**) have $\alpha = 10^{-5}$ and 10^{-3} and damping $g = 10^{-2}$. The bottom two (**ii** and **iv**) have $\alpha = 10^{-5}$ and 10^{-3} and damping $g = 10^{-1}$. D_0 is the coefficient of the gradient term in the free energy of (1.18): $D_0(\nabla P)^2/2$. It is related to α as $\alpha = A_0^2 D/m \approx 10^{-3}$, where A_0 is coefficient of the quadratic term in the free energy of (1.2): $A_0(T - T_C)P^2$ and is approximately 6×10^{-6} K^{-1}; and the parameter m is about 6.4×10^{-21} kg m^{3}s^{-2}A^{-2}. Parameter g is the dimensionless ratio of damping constant to undamped phonon frequency γ/ω_0

1.2.1 Small Particles

The situation with regard to small ferroelectric particles has been even more confusing and misleading than that with thin films. Many early studies [48], particularly those of Uchino [49], incorrectly interpreted decreases in polarization with decreasing particle size as true intrinsic size effects due to depolarization fields, occurring at diameters as large as 100 nm. We now know that this is false, that they were due to surface tension, strain, increase in porosity [50], or other pedestrian effects; intrinsic size effects do not occur until around 15 nm, as has been independently and carefully demonstrated by Payne [51], Cao and Randall [51], and especially Tanaka and Makino [52]. The latter two authors show that barium titanate particles consist of a shell of cubic material surrounding a core of tetragonal materal; as the particle sizes become less than 20 nm, the core vanishes. But this has nothing whatsoever to do with depolarization fields; it is material-specific and not a generic effect. Although most nanoscale ferroelectric studies have been on barium titanate, PZT-family nanoscale powders 14–60 nm in diameter have been prepared via sol–gel techniques by Tang et al. [52].

2. Basic Properties of RAMs (Random Access Memories)

As of 1988 there were fourteen generically different kinds of digital electronic memories in use (Fig. 2.1a). These ranged from the slow but cheap and high-density disks for archival storage, to the fast Static RAMs (SRAMs). It was anticipated [53] that the memory market would simplify greatly over the years 1988–98 by the introduction of ferroelectric NonVolatile (NV) RAMs, eliminating core, magnetic bubble memory, and other devices (Fig. 2.1b). The aim of this chapter is to review that development and its present status. In the 1980s much of the motivation (and funding) for the development of ferroelectric NV-RAMs came from the US military. The reasons are illustrated rather graphically in Fig. 2.2. As shown, a typical military aircraft has 96 kbit of EEPROM in its embedded radar warning receivers, 8 Mbit of EEPROM or bubble memory in its removable cockpit recorder, 96 kbit of EEPROM in its electronic counter-measures (and counter-counter measures) receivers, 36 Mbit of SRAM or EEPROM in its main mission computer, 48 kbit in its EEPROM missile memory, and another 200 Mbit of bubble memory or EEPROM for its flight incident recorder and secondary memory. However, by 1999, this impetus has decreased dramatically. The US military has 'no enemies' with radiation-producing weapons requiring radiation-hard memories in its military aircraft; and the interest in ferroelectric RAMs has shifted strongly to the commercial sector (including commercial satellites). The future for ferroelectric RAMs (Fig. 2.1c) is in smart cards, audio memories in SONY Walkman players, in toaster-ovens, and in personal computers, not in guided missile memories – and perhaps that is all for the better.

Panasonic and NEC have kindly provided figures and data for the present chapter, much of it unpublished, on their devices. Therefore these are emphasized below. Similarly, much of Chap. 17 is from SONY. However, I believe that these are representative of the work in the industry, and that other corporations, including Ramtron, Symetrix, Toshiba, Hitachi, Fujitsu, Samsung, Siemens, etc., have comparable data, devices, and innovations, and this section is not intended as advertising for any specific corporation.

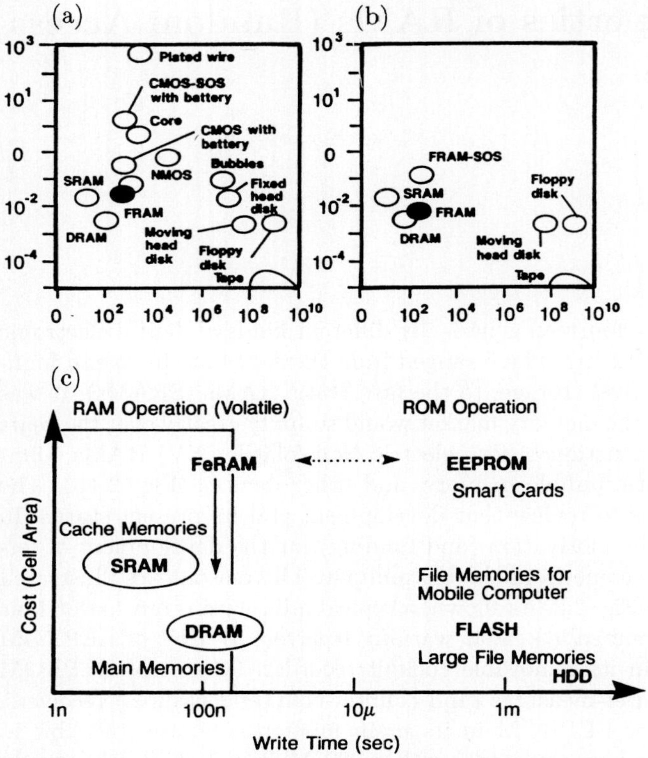

Fig. 2.1. Cost per bit versus access time: (a) for the 14 kinds of digital memory in use in 1988 [53]; (b) 1988 projection to 1998; (c) 1999 comparison (NEC)

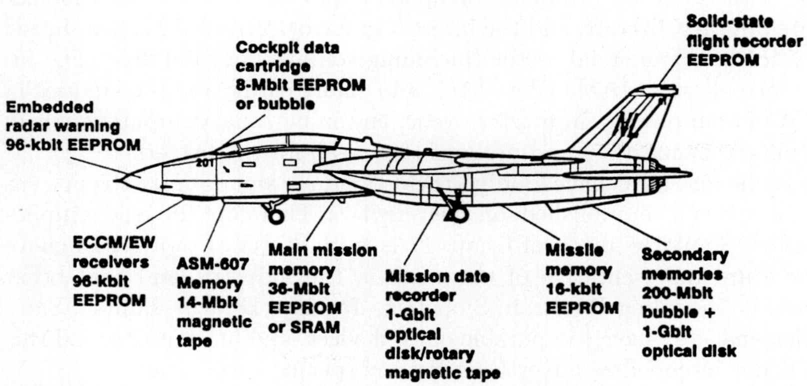

Fig. 2.2. Nonvolatile memories in a US military aircraft (1988); the original figure is from the North American Defence Center (NADC), and its reproduction in [53] was through the private communication of R. Fedorak

2.1 Schematic Designs

Although this text is intended to be useful for integrated circuit engineers, solid state physicists, physical chemists, and ceramists, only the engineers will be familiar with the jargon which follows: bit-lines, word-lines, V_{CC}, etc. Hence the present section is a brief primer for the novice to illustrate the basic ideas in a ferroelectric random access memory.

Such memory devices can in principle be designed as JFET (Junction Field Effect Transistors), bipolar, or MOSFET (Metal Oxide Silicon Field Effect Transistor) designs. In fact, in the early 1990s McDonnell-Douglas designed ferroelectric JFET memories; and Raytheon made bipolar prototype ferroelectric nonvolatile RAMs. However, at present almost all designs are single-carrier devices (MOSFETs) and generally n-channel (enhancement) MOSFETs.

A possible source of confusion in the recent literature are references to SRAM (Static RAM) designs for ferroelectric memories. No ferroelectric memories at present are actually SRAMs. However, a normal SRAM layout is a flip-flop circuit used to store a single bit of information. It has six MOSFETs per cell: four enhancement mode MOSFETs and two depletion-mode MOSFETs (it is sometimes possible to replace the two depletion-mode MOSFETs by polysilicon resistors). The ferroelectric memories include so-called 2T–2C devices with two MOSFET transitors and two ferroelectric capacitors in a geometry which resembles the 4-MOSFET SRAM, so the 'SRAM' label is occasionally used.

Before going on to examine various design subtleties in this chapter, let us remind readers of the basic notation and circuitry of simple bipolar devices and MOSFETs. Then we shall proceed to show how the ferroelectric IC devices derive from them directly.

Figure 2.3a below shows the basic structure of a simple bipolar p-n-p device. It consists of a p-type emitter (heavily doped with acceptors to make it nearly metallic, this is denoted p+), an n-type base, and a p-type collector (the collector has a carrier concentration of about 10^{16} cm^{-3} in Fig. 2.3a, compared with say 10^{19} cm^{-3} for the p+ emitter). The device shown is configured in a common-emitter configuration, which means that the emitter lead is common to both input and output circuits. (A frequent alternative configuration is the common-base layout.)

The operation of such a circuit has four possible modes, depending upon the signs and magnitudes of the voltages applied between the emitter and base (V_{EB}) and the collector and base (V_{CB}). Note that the applied voltage V_{CC} differs from the collector–base voltage V_{CB} due to the load resistance R_L. The modes are designated as follows: (a) the saturation mode has both V_{EB} and V_{CC} forward biased (> 0); (b) the active mode has V_{EB} forward biased but V_{CC} reverse biased; (c) the cutoff mode has both voltages reverse biased; and (d) the inverted mode has V_{EB} in reverse bias and V_{CC} forward biased. When FETs are used for analog devices, only the forward biased *active* mode is employed, but in digital circuits, all four modes are used. When a fast-rise-time square voltage pulse is applied to take V_{EB} or V_{CC} from positive to negative, or vice versa, this n-channel MOSFET acts like a very fast (> 1 GHz) switch. The collector current I_{C} changes rapidly and by orders of magnitude.

The output characteristics of this p-n-p transistor in the common-emitter configuration is shown in Fig. 2.3b. For each value of base current I_{B} ($\approx$ 20 μA) there is a collector current I_{C} of order 4 mA (a gain of around 200). The saturation, active, and cutoff regions are shown in the figure (the inverted mode is similar to the active mode, except that the currents are smaller due to a less efficient p-type 'emitter' than the p+ used in the active mode). Figure 2.3b superimposes on this set of $I(V)$ curves a so-called *load line*, which is simply determined by the applied voltage V_{CC} divided by the load resistance R_{L}. The operating point of the device is where the lines cross for a chosen dc base current I_{B}.

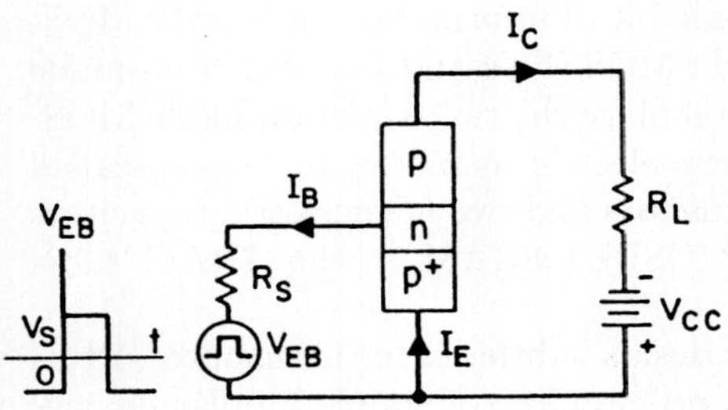

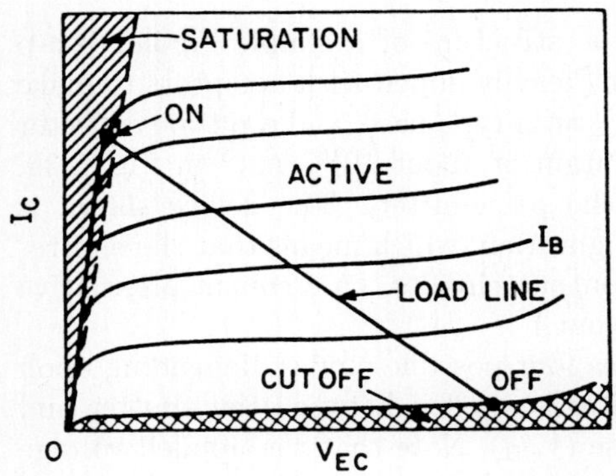

Fig. 2.3. *Top*: common-emitter p-n-p transistor schematic. *Bottom*: output characteristics

Now to extend this basic idea to a ferroelectric RAM is rather simple. As shown in Fig. 2.4 the application of an applied voltage to the ferroelectric capacitor gives both a charge Q and hence a current I_{C}, just as in Fig. 2.3b. Only the shape is different in two respects from the p-n-p transistor current I_{C}: firstly, it saturates less abruptly; and secondly, it consists of two curves

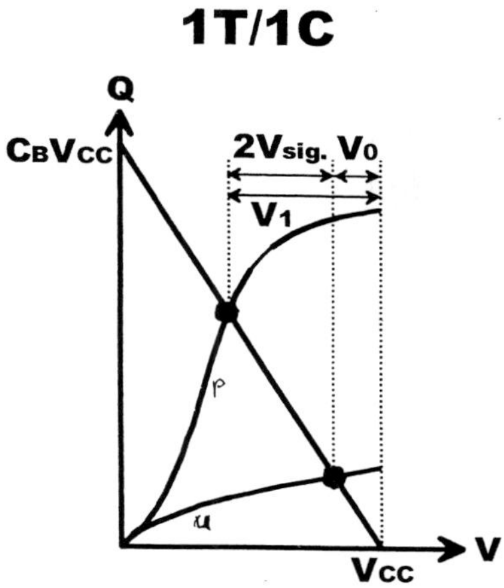

Fig. 2.4. Load line superimposed on ferroelectric hysteresis curves (P and U of Fig. 1.1), showing operation points. Compare with the p-n-p common-emitter circuit in Fig. 2.3

(the upper and lower parts of the hysteresis loop, labeled P and U in Chap. 1, Fig. 1.1. In Fig. 2.4 we again superimpose the load line, and we observe that there are two operational points for this device, corresponding to a 1 and to a 0 (upper and lower) in the memory. As before, the load line is linear with slope $-1/R_L$ and intercept V_{CC}.

Figures 2.5 and 2.6 illustrate the WRITE operation and READ operation, respectively, from this picture. These figures replace the basic p-n-p transistor terminology of emitter, base, and collector with the IC jargon word-line (WL), bit-line(BL), and plate (PL), but the ideas remain the same. (A practical complication of ferroelectric memory construction is that the plate must physically also be the electrode of the ferroelectric capacitor; thus, its area may need to be larger – several square microns – than would normally be the case in an Si device, in order to provide enough switched charge for the sense amplifiers.) In the WRITE operation (Fig. 2.5) the voltage on the plate is suddenly pulsed from zero to $+V$ or $-V$. The $+V$ results in driving the ferroelectric to its $-P_r$ remanent polarization state (a 0 in the memory), and $-V$ on the plate drives it into $+P_r$ (a 1 in the memory).

From these simple figures the design engineer can extract quite a lot of quantitative information for ferroelectric RAMs. For example, as shown in Fig. 2.7, 'imprint' is a preferred state of internal voltage bias in the ferroelectric capacitor. In it the hysteresis curve may shift to the $+V$ side of zero, as shown by the inner two curves, from the neutral value shown by the outer two curves. When this occurs, the operating points from the load line cross the upper and lower branches of the hysteresis curves at different points – points that are closer together than initially in the neutral state. This leads to discrimination problems for the sense amplifiers in distinguishing 1 from 0 in the memory. Fatigue can be diagrammed with a very similar set of diagrams. In each case, it is the position of the load line through the modified hysteresis curve lines that gives the desired information at a glance.

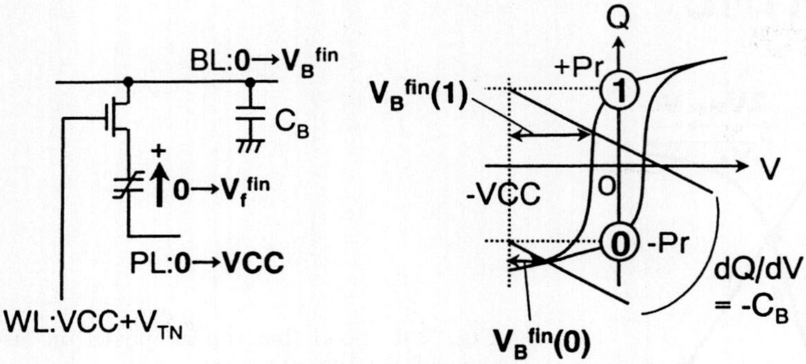

Fig. 2.5. (*Left*) FRAM WRITE operation (schematic). (*Right*) hysteresis

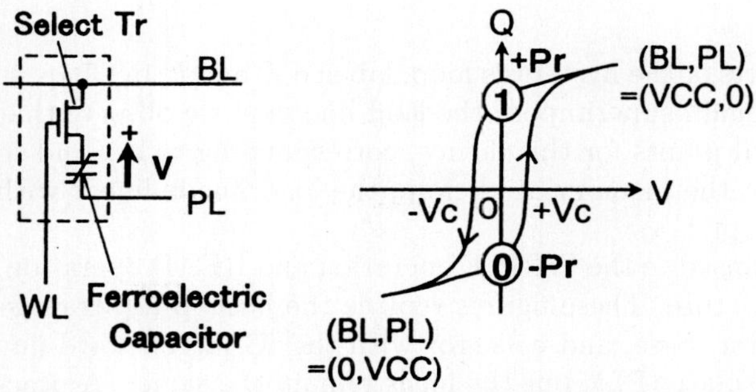

Fig. 2.6. (*Left*) FRAM READ operation (schematic). (*Right*) hysteresis

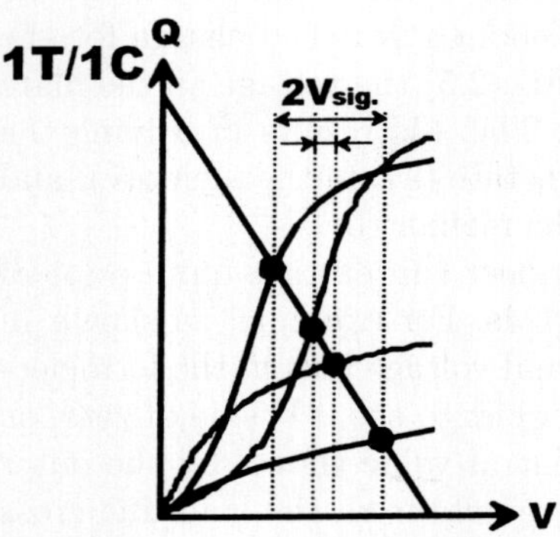

Fig. 2.7. Imprint in a 1T/1C cell

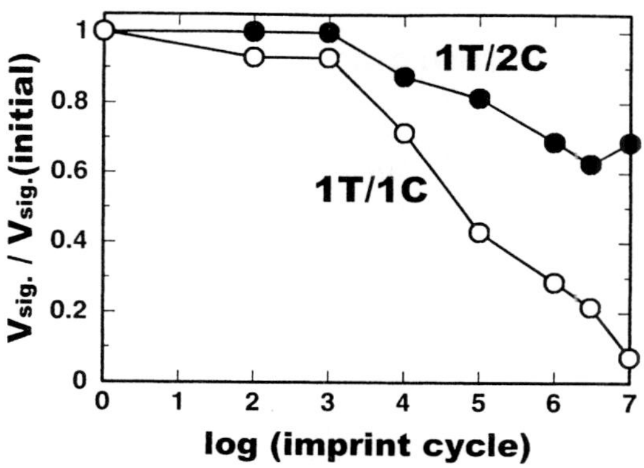

Fig. 2.8. Imprint NEC 8 kbit TEG

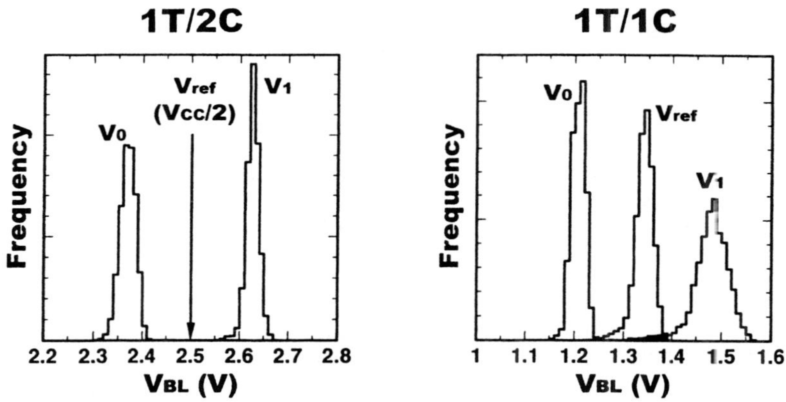

Fig. 2.9. NEC test values of bit-line voltage distribution for 1T/2C and 1T/1C cells

Note that the difference in voltage between the two operational points in Fig. 2.7 is defined as twice the signal voltage $2V_{\mathrm{sig}}$. The signal voltage is one of the key design parameters. In an electrical testing of a ferroelectric RAM, it is actually the degradation in READ signal voltage (and not P_{r}) that is measured. This is shown in Fig. 2.8 from NEC for their 8 kbit PZT test chip (TEG). Similarly, the bit-line voltage V_{BL} is another key characterization parameter. Comparison of the distribution of measured V_{BL} values for both 1T/1C and 1T/2C cells is shown in Fig. 2.9. Here V_1 is the difference between V_{CC} and the first (1) operation point; V_0 is the difference between V_{CC} and the second (0) operation point. Note that the spread of V_1 values is greater than that for V_0 in a 1T/1C cell, whereas they are the same in the 1T/2C cell. This is an engineering advantage of the NEC 1T/2C cell design.

The way V_{BL} is actually measured is shown in Fig. 2.10. The bit-line voltage after the sense amplifiers is monitored as the input voltage V_{in} is

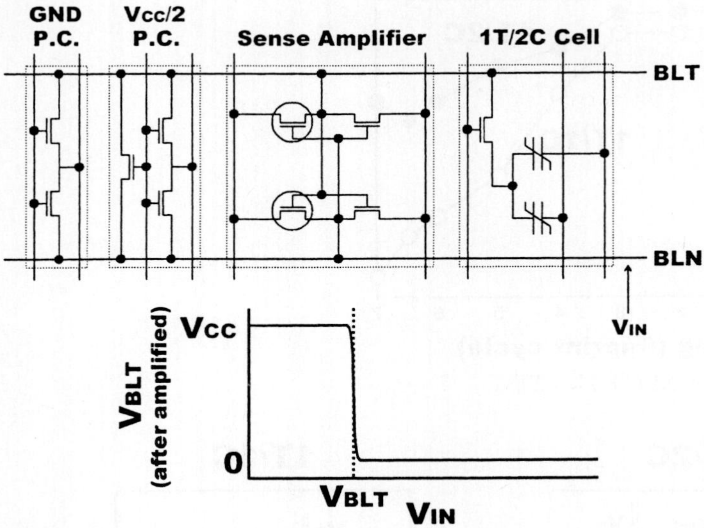

Fig. 2.10. V_{BL} measurement (NEC)

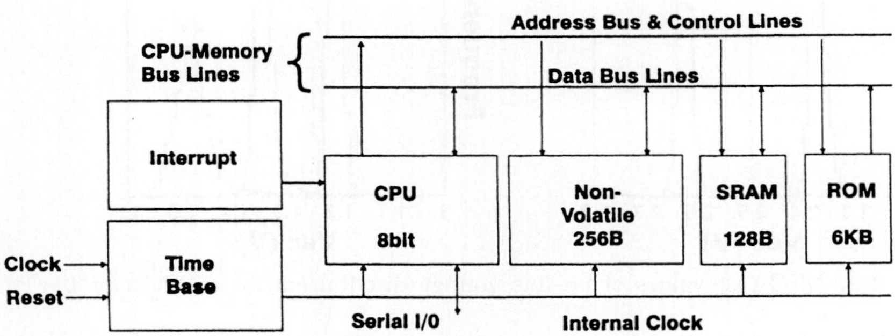

Fig. 2.11. NEC FeRAM module (1998)

ramped up. V_{BL} drops abruptly from V_{CC} to near zero at $V_{\mathrm{in}} = V_{\mathrm{BL}}$ and is monitored as the input voltage on the bit-line is increased.

NEC has combined a small 256 bit ferroelectric RAM with an 8 bit CPU, 128 bit SRAM, and 6 kb of ROM (with internal clocks, etc.) into an FeRAM module (Fig. 2.11). This small modular element may soon make it easy for engineers to design many systems for inexpensive products. I applaud the effort to modularize these units and not just concentrate on very large arrays. Engineering progress is often at the bottom end and not always at the top end.

There are a number of different electrical engineering designs of ferroelectric memories. We outlined above the most popular devices utilizing one or two transistors and one or two ferroelectric capacitors per cell (i.e., per bit stored). Not included were shadow RAMs. Shadow RAMs are random access memories that operate in a fast memory system but dump their stored information in a nonvolatile 'save' memory in certain events, such as power interruption. A ferroelectric shadow RAM does not seem to be a viable commercial product, but some were described in the literature from 1988–92 by companies such as Ramtron (Fig. 2.12).

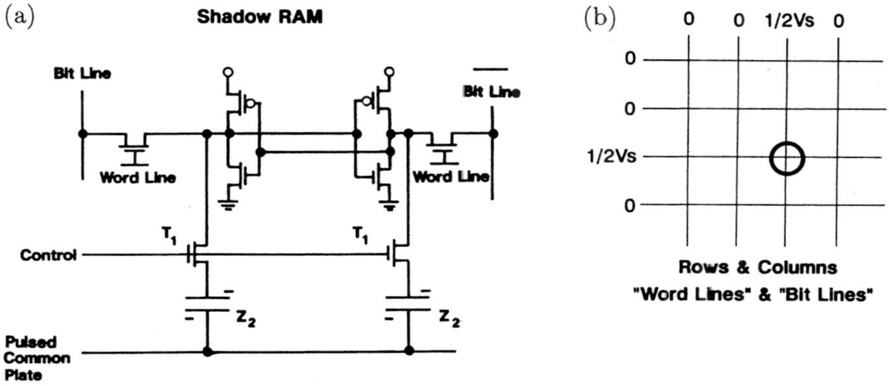

Fig. 2.12. (a) Shadow RAM (Ramtron). (b) Raw-array random access memory

Note that the simplest kind of random access memory, illustrated in Fig. 2.13, cannot be used. This raw array of memory cells consists of rows and columns (*bit lines* and *word lines*) with electrical interconnects that permit half the requisite switching voltage to pass as a short voltage pulse along a row, and half along a column. The two $V/2$ pulses intersect simultaneously at the bit to be addressed or 'selected'. Under ideal circumstances in which the material in the memory cell has a true voltage threshold, none of the other bits will switch. But unfortunately, ferroelectrics never have a true threshold. Instead, at modest fields their switching follows an activation field dependence, with the switching probability over a given time t_s given by

$$t_s = t_0 \exp(-E/\alpha) , \tag{2.1}$$

where the activation field α is typically of order 100 kV/cm (10 MV/m). Equation (2.1) is confirmed nicely for KNO_3 in Fig. 2.13. Earlier work [54, 55] had suggested that KNO_3 had a true threshold voltage (or field) for switching, but we know now that this is untrue [56, 57]. This activation field dependence means that the $V/2$ half-select row voltage or column voltage can occasionally disturb or unintentionally switch or 'select' a bit on the same row as the addressed bit, producing a bit-error in the memory (a 1 where a 0 should be). In a typical ferroelectric, this half-select disturb pulse might cause an error rate of 10^{-4}. Therefore a 1 Mbit chip would have an unacceptable error rate

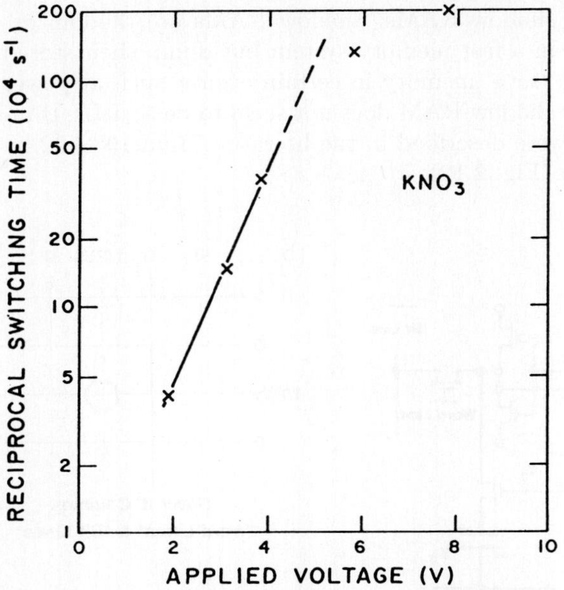

Fig. 2.13. Exponential activation-field dependence of switching time in potassium nitrate [57]

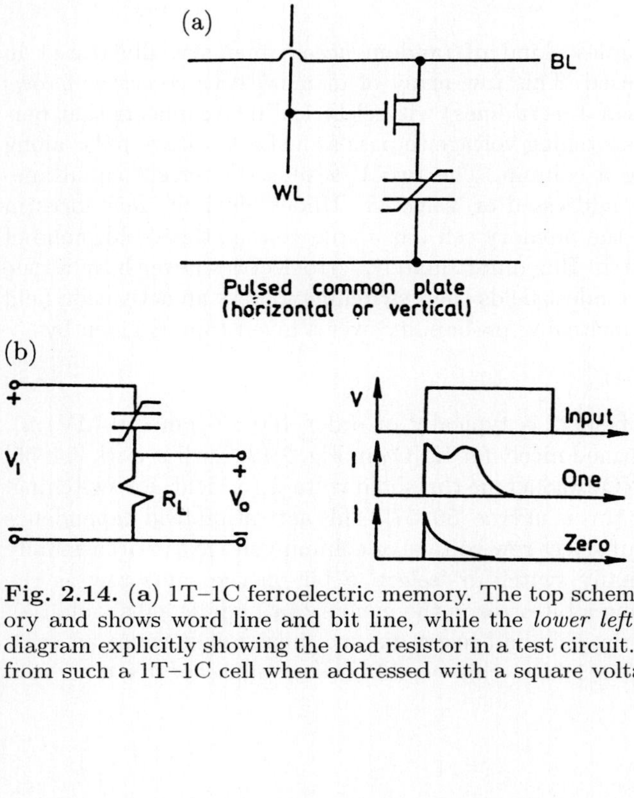

Fig. 2.14. (a) 1T–1C ferroelectric memory. The top schematic is for a ULSI memory and shows word line and bit line, while the *lower left figure* is an alternative diagram explicitly showing the load resistor in a test circuit. (b) The current output from such a 1T–1C cell when addressed with a square voltage pulse

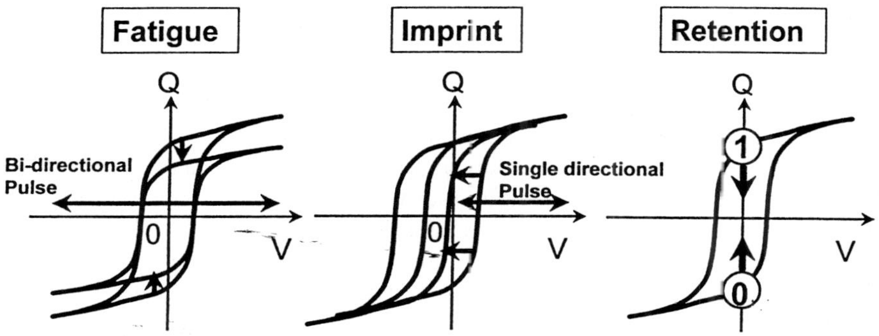

Fig. 2.15. Effects of fatigue, imprint, and loss of retention in 1T/1C cells

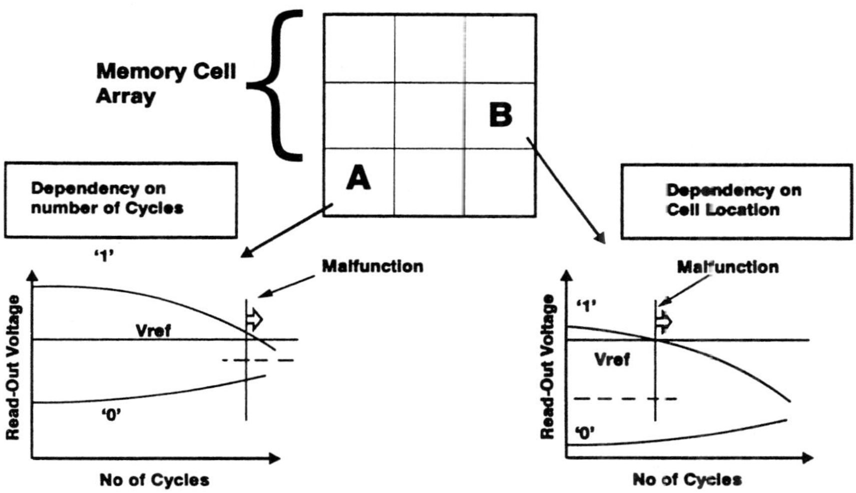

Fig. 2.16. Problems for 1T–1C memories

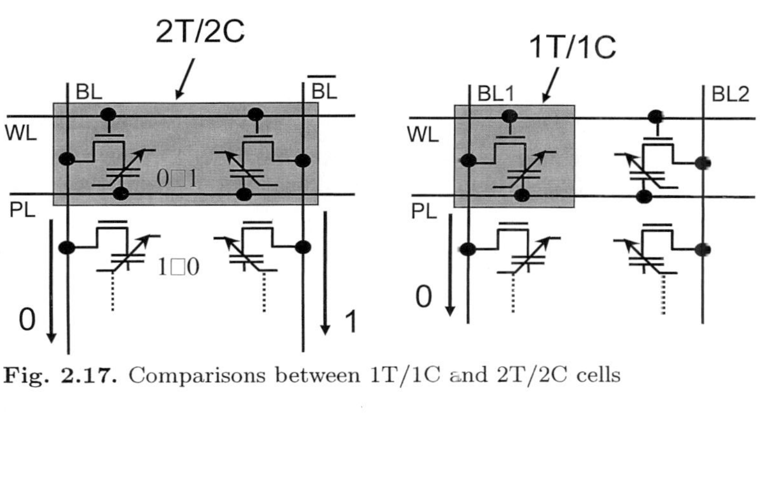

Fig. 2.17. Comparisons between 1T/1C and 2T/2C cells

of 100 errors each time the memory was written. In order to make an error-free ferroelectric RAM it is necessary to put a pass-gate transistor behind each ferroelectric cell, so that the transistor provides the sharp threshold in switching voltage and eliminates all switching from $V/2$ half-select pulses.

The simplest structure, involving one transistor per cell and one capacitor (1T–1C design) is shown again in Fig. 2.14a. Note that in this design the READ operation is not simple: one must apply a switching voltage to the cell (bit), measure its output current through a small (< 10 ohm) load resistor, send this current to a sense amplifier to see whether it is a large displacement current due to polarization reversal (curve P in Fig. 1.1), or a smaller current due to the linear response of a capacitor already polarized in the same sense as the applied voltage (curve U in Fig. 1.1). (The sense amplifier does this by subtracting the linear response U of a reference capacitor that is always polarized positive.) Then after each READ, which destroys the original polarization sense, a reset voltage must be applied to restore the original polarization. This destructive READ operation and its associated restore pulse are a source of fatigue in ferroelectric memories. There are other problems with 1T–1C cells: fatigue (decrease in P_r with bipolar cycling), imprint (development of a bias field) and retention (decrease in P_r with time) are illustrated schematically in Fig. 2.15. Rather more subtle, and shown in Fig. 2.16, is how the reference voltage for the sense amplifier depends upon cell location in the array. This militates against a single reference cell in the array and mitigates in favour of a 2T–2C cell structure where each bit has its own reference (Fig. 2.17).

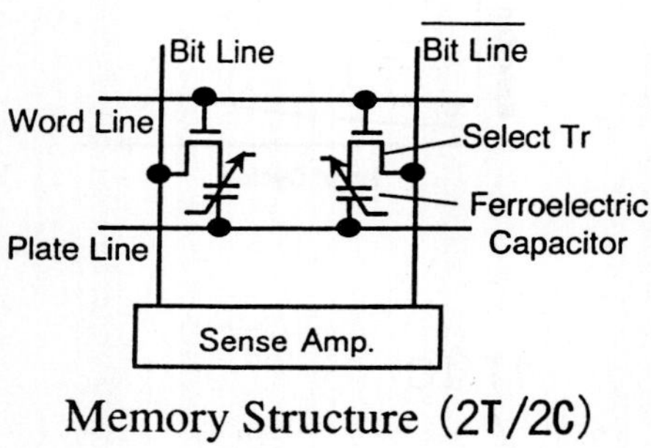

Fig. 2.18. 2T–2C ferroelectric memory

The reference cell with which the sense amplifier compares output voltages can be a single cell in the entire memory (1T–1C circuit described above), or it can be a reference cell in each bit, as shown in Fig. 2.18. The circuit shown in Fig. 2.18 is called 2T–2C and has two transistors and two identical ferroelectric capacitors for each bit. While the 2T–2C circuit uses up much more space than the 1T–1C and is therefore unsuitable for very high bit-density NV-RAMs, it has the advantage that the reference cell fatigues at the same rate as the addressed cell, so that the sense amplifier always gives a null voltage when a 0 is sensed. The READ operation in a 2T–2C memory cell is carried out as diagrammed in Fig. 2.19. Here the signal charge density q/A must be as shown in the figure and is defined through a ratio

$$K_s = \frac{q}{2AP_r} . \tag{2.2}$$

Typically there will be a 500 fC capacitor and it must drive a bit-line signal with a 100 mV sense amplifier margin. As shown in Fig. 2.20, the ratio K_s is almost an order of magnitude better than in PZT. This means SBT NV-RAMs will be 10 times easier to read than PZT. The actual ratios K_s are 0.41 at 2 V, 0.61 at 3 V and 0.66 at 5 V for SBT; 0.07 at 2 V, 0.09 at 3 V and 0.18 at 5 V for PZT. In my opinion, a PZT NV-FRAM will never operate at the intended 1.1 V Si logic level. (It will also be difficult to make flash EEPROMs operate at 1.1 V. This is a strong advantage of FeRAMs over flash EEPROMs, perhaps the greatest advantage.) The emphasis in Table 2.1 is upon comparisons between FeRAMs and flash EEPROMs. This is apt to be of greatest commercial importance in the next five years.

Table 2.1. Parameters of nonvolatile memories

	Bits	Access [ns]	Rewrite time[a]	Cell size (L = design rule)[b]	Voltage [V]
SNOS	8k × 8	35	11 ms		
MNOS	1 Mb	150	10 ms		
SONOS	8k × 8	150	10 ms		
Magnetic RAM	16k × 1	200–2000	200 ns		
FeRAM	1 Mb	60	60 ns	$4L$	1.1
EEPROM (floating gate)	1 Mb	150	10 ms	Large	1.1 ?
EEPROM (flash)	1 Mb	120	10 ms	$3.2L$	4.5–5.0
Core	128k	350	900 ns		

[a] Per byte for SNOS, MNOS, SONOS, EEPROMs.
[b] Hence cost.

NEC has developed an intermediate circuit for prototype NV-RAMs involving one transistor and two capacitors per bit. This 1T–2C cell is shown in Fig. 2.21 and is used at present in PZT devices.

Most large memory companies have well defined corporate strategies for the next ten years of ferroelectric chip development. That for NEC is shown in Fig. 2.22, and for Panasonic in Fig. 2.23. In each case the initial goal

Table 2.2. Comparison between FeRAM and other memories

	FeRAM	DRAM	SRAM	FLASH
Nonvolatility	○	×	×	○
Write cycle	10^{6-12}?	$> 10^{15}$	$> 10^{15}$	10^{5-6}
Write speed	$\sim$ 100 ns	$\sim$ 50 ns	$<$ 10 ns	$\sim$ 1 ms
Read cycle	10^{6-12}?	$> 10^{15}$	$> 10^{15}$	$> 10^{15}$
Standby	$<$ 1 μA	$>$ 10 μA	$>$ 1 μA	$<$ 1μA
Operation	RAM	RAM	RAM	ROM
Cell size	(1)	1	4	0.8

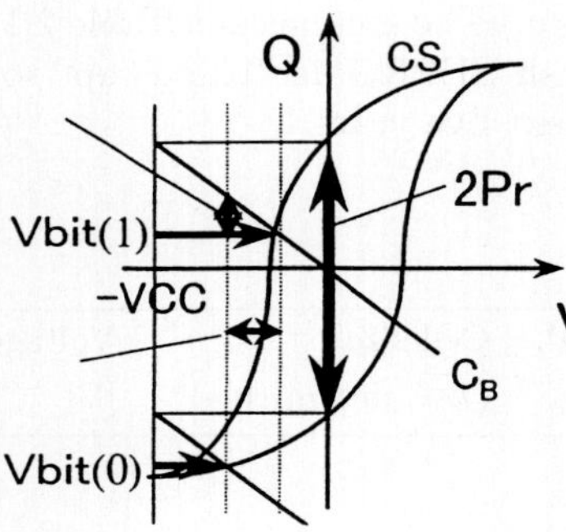

Fig. 2.19. READ operation in a 2T–2C memory. Small *double-headed arrows* point to the signal charge and the signal voltage

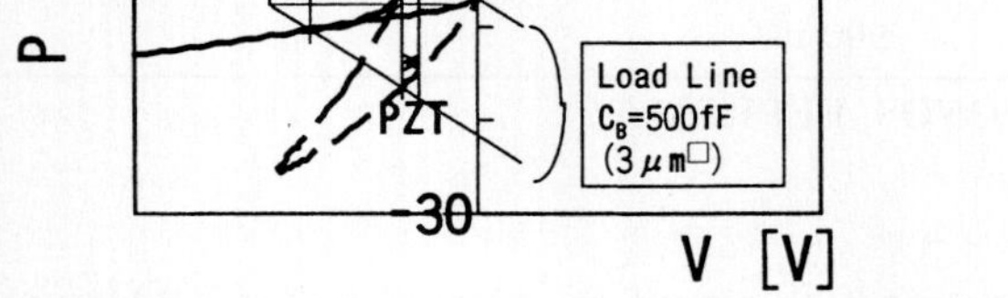

Fig. 2.20. Ratio K of charge to $2P_r$ in PZT and SBT. Hysteresis curves for PZT and SBT showing better K ratio in SBT at the same P_r value

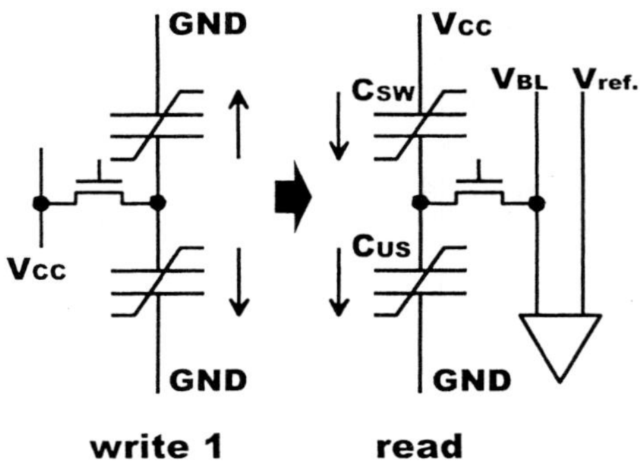

Fig. 2.21. 1T–2C ferroelectric memory

is replacement of EEPROM for IC (Integrated Circuit) cards and microcontrollers, followed in the years 2000–2005 by replacement of flash, DRAM, and SRAM in the Mbit–Gbit regime. However, the low-density end of the market is quite important. Figure 2.24 shows projected sales of IC cards exceeding \$3 thousand million/year by 2001.

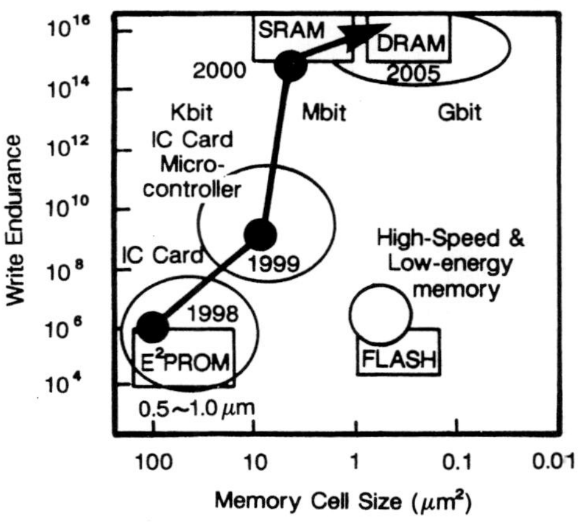

Fig. 2.22. NEC corporate plan for FeRAMS (1998)

The application to smart cards (IC cards with central processing unit – CPU – core memory) has as an advantage that memory access from the READ/WRITE operation is permitted only through the CPU in a smart card Large Scale Integrated circuit (LSI). Thus the CPU program provides the smart card with added security. In addition, the co-processor can provide additional encryption and this results in a very high security card.

Applications of these smart card LSIs will include global satellite mobile phones, automatic tolling for highways, authentication for computer networks, Medicare and other national health cards, selective reception of com-

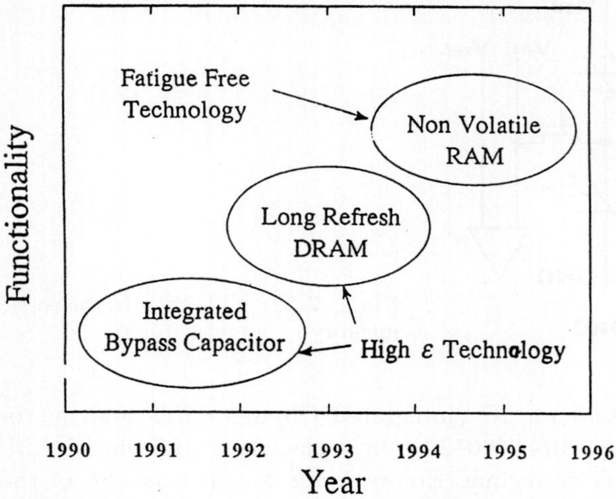

Fig. 2.23. Panasonic corporate plan for FeRAMS (1990)

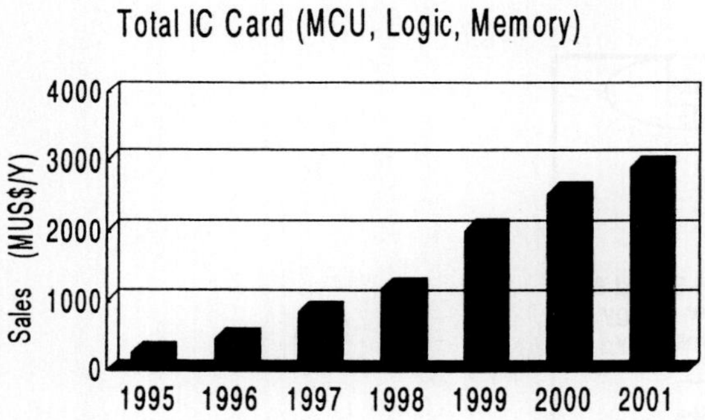

Fig. 2.24. Projected sales for IC cards (US$/year) from NEC

munications satellite digital broadcasts, train tickets, and authentication for electronic conferencing.

The conventional smart card LSI with an EEPROM memory has the disadvantage of a very low-speed WRITE operation, and a WRITE endurance of $< 10^6$ times. Its advantage is that it has a small chip area, many available circuit libraries, and an up-to-date CMOS process for its CPU. Another disadvantage is that it requires a multilayer metal process. The conventional smart card LSI with a FeRAM has a much faster WRITE speed and much greater WRITE cycle endurance. It also works with a simpler single-layer metallization. However, it requires a different processing for its CPU (with lower performance).

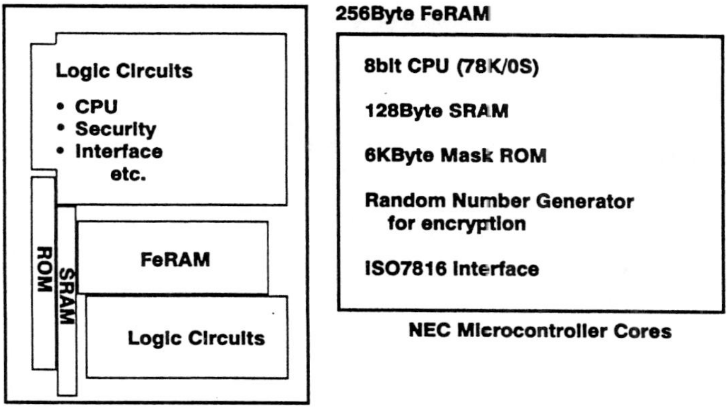

Fig. 2.25. NEC smart card parameters. This is a 0.8 μm double-Al layer CMOS

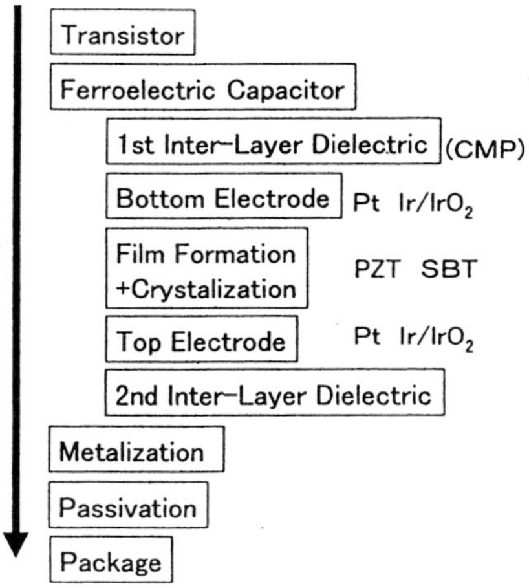

Fig. 2.26. Typical processing steps for ferroelectric memories using PZT or SBT and Pt or Ir electrodes

NEC has recently developed a smart card LSI with the parameters shown in Fig. 2.25. This has a 0.8 micron double layer aluminium CMOS process. It operates at 5 MHz at 3 V. The FeRAM unit is modularized, so that it can be put, whole-cloth, into other devices.

The processing of these devices is discussed in more detail in Chap. 11, but Fig. 2.26 shows the general process sequence used to fabricate these devices, and Fig. 2.27 illustrates the process integration. The selection of ferroelectric material at NEC is still moot. Figure 2.28 shows that remanent polarization is satisfactory in both PZT and SBT but that there is unexpected breakdown in some SBT films at rather low voltages (about 2 V). This is not understood and may arise from very pedestrian causes, such as irregularities in film thickness. Finally, Fig. 2.29 shows that iridium/iridium oxide electrodes are superior to Pt for PZT devices, at least with respect to fatigue [58].

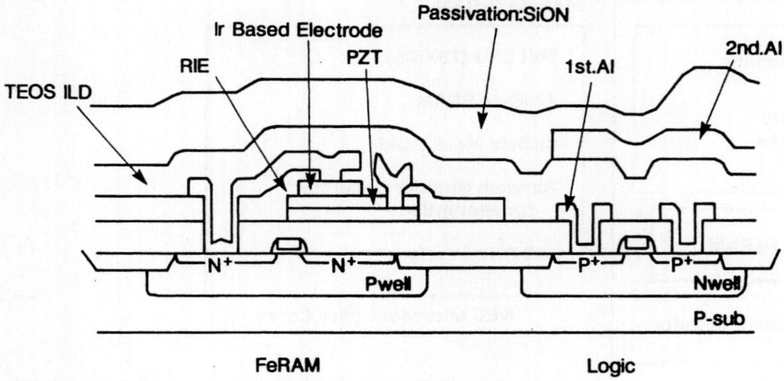

Fig. 2.27. Cross-section for NEC PZT memory

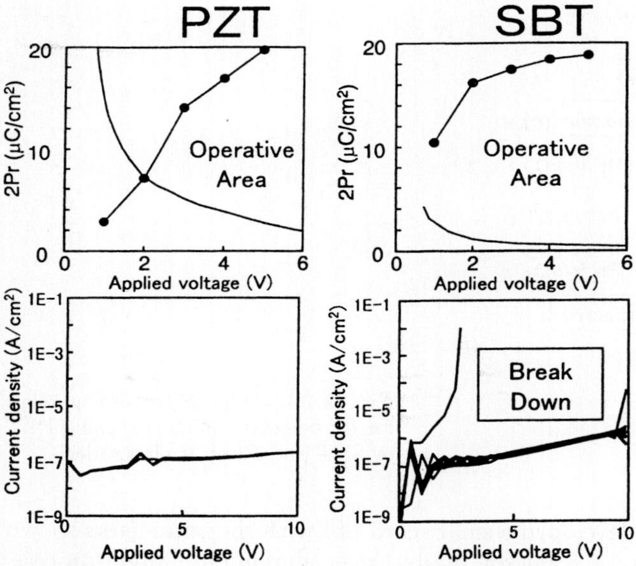

Fig. 2.28. Remanent polarisation and leakage current versus voltage for PZT and SBT from NEC

The other technological developments of importance at NEC, in addition to a reliable double-metal CMOS process, include a screening circuit that uses an offset sense amplifier to increase overall chip reliability by detecting weak cells in an initial test. This is diagrammed in Fig. 2.30. There is also (Fig. 2.31) a self-referencing READ scheme, in which each memory cell generates its own reference voltage by accessing the same cell twice. The sense amplifier has a new latch type (Fig. 2.32) that permits the bit-line voltage to be stored temporarily.

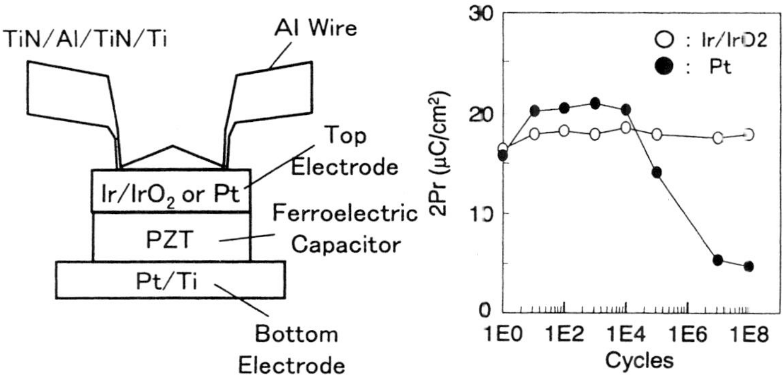

Fig. 2.29. Structure of NEC PZT memory and fatigue with Ir or Pt electrodes

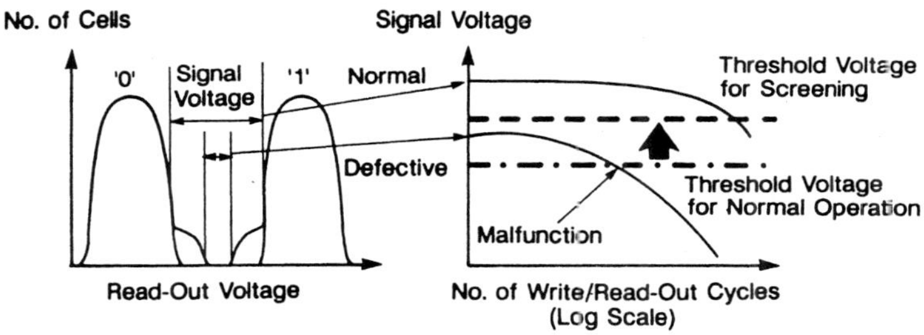

Fig. 2.30. Number of cells with given readout voltage, and signal voltage versus write cycles for PZT (NEC)

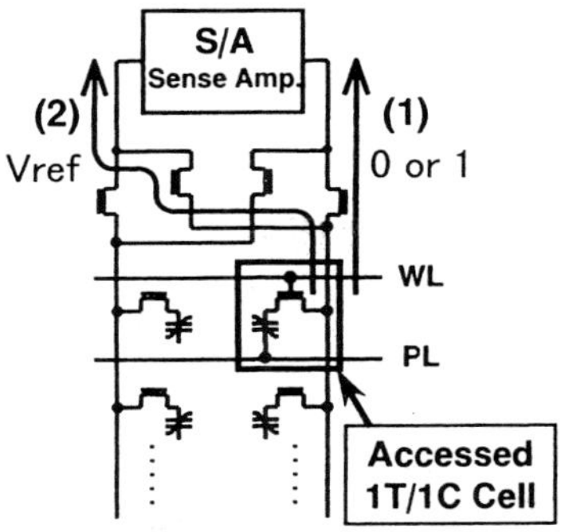

Fig. 2.31. Self-referencing FeRAM cell (NEC)

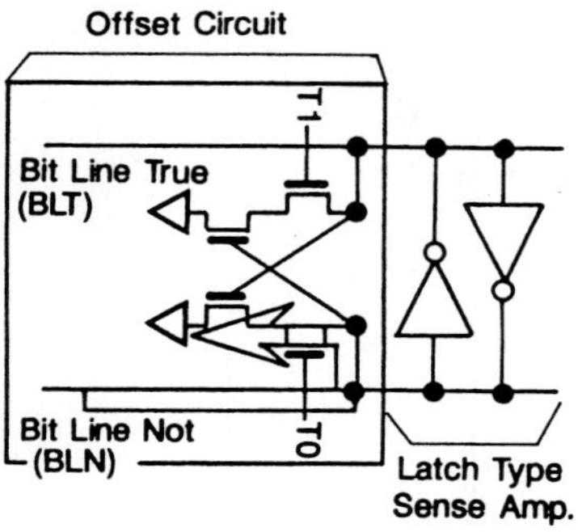

Fig. 2.32. Latch structure for sense amplifier in NEC NV-RAMs. The switch transistor (*arrow*) makes the bit line (BLN, BLT) floating and enables the bit-line voltage to be stored temporarily

2.2 Actual Devices

Ferroelectric memories are already (April 1999) in the kbit class, used in smart cards and as substitutes for EEPROMs in general [59, 60, 61]. Within a year or two they will serve as general purpose memories in the Mbit class [62] replacing SRAM, DRAM, and flash memories. Table 2.1 compares a variety of nonvolatile memories, and Table 2.2 gives a somewhat different comparison among FeRAMs, DRAMs, SRAMs, and flash.

At present there are a number of ferroelectric memories in at least limited commercial production. These range [59, 60, 61] from low density 4 or 64 kbit smart card chips (Fig. 2.33) to the 256 byte, 256 kbit (Fig. 2.34), and 1 Mbit free-standing RAMs [62]. The Panasonic 256 kb FeRAM is a 32k × 8 device with 1.2 micron CMOS (Complementary Metal Oxide Semiconductor) fabrication and single-level aluminum metallization. The cells are 11.4 × 7.5 microns, with 3 mA active current and 0.1 mA standby. The RC time constant is 200 ns. The cell area has 4 blocks, each of which contains 64 kbit. Row decoders and word drivers are between pairs of blocks, and column decoders and sense amplifiers are on the right side of the array. Address buffers, row decoders, and wordline boost circuits are on the left. I/O buffers, clock generators and column address buffers are on the extreme right. Note that all these devices are vertically integrated with the FE-film directly above the Si RAM; this contrasts with earlier devices, such as the FE-JFET device (Junction Field Effect Transistor) made on GaAs (Fig. 2.35), which the FE capacitor sits adjacent to the GaAs JFET; such a design was a good R&D prototype but cannot be scaled to high densities. Note also that the FE-GaAs JFET was not planarized. The geometries of the new 1 Mbit FeRAMs are comparable in size with state-of-the-art electroding; 200 nm Pt post electrodes from Tegal are shown in Fig. 2.36, which may be compared with standard Pt electrodes from National/Panasonic in Fig. 2.37.

Of special commercial importance are fully embedded devices (Fig. 2.38a and b) in which a ferroelectric NV-RAM is embedded in a Si microcontroller in a single chip. These embedded devices can be used in household electrical

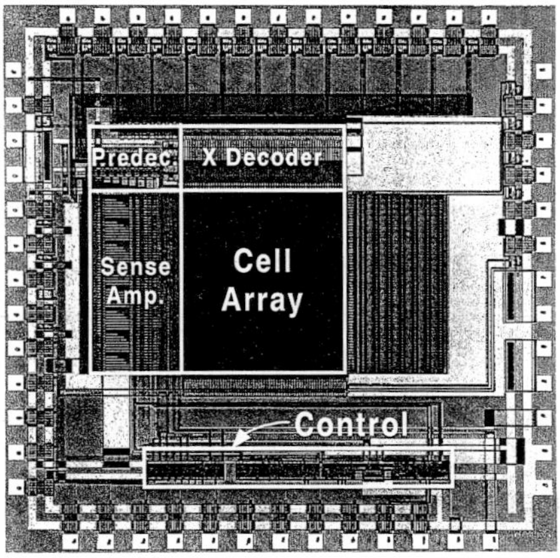

Fig. 2.33. NV-FRAM smart card (debit card) 16 kbit from NEC. This is a 0.8 μm double-Al device with a 9 × 8 μm² cell, and a 3.7 × 3.7 mm² chip (5 V) with 6.3 mW dissipation and 1.8 μs cycle time

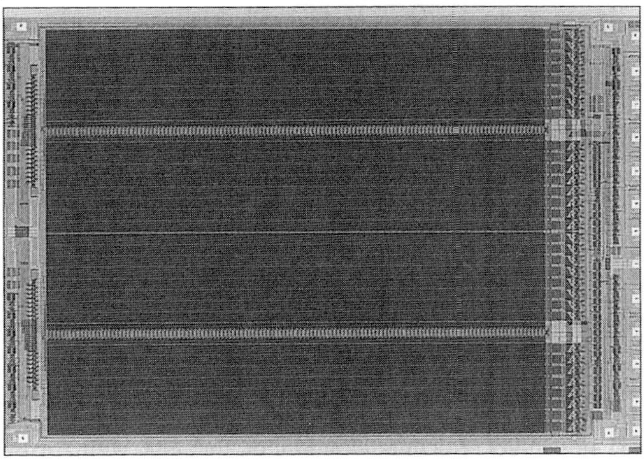

Fig. 2.34. 256 kbit SBT NV-RAM (Matsushita–Panasonic) [59, 60, 61]

appliances, such as washing machines, dishwashers, dryers, toaster ovens, blenders, etc., in which the bit density requirements are modest. This is a large volume market.

As soon as these embedded ferroelectric NV-RAMs can be produced in large volume at the 4 Mbit level, they should find application in digital cameras. At 256 Mbit they can be used in audio memories, such as the SONY Walkman. The main competition will be from flash EEPROMs. Flash devices can be made slightly more dense than ferroelectric RAMs (area per bit is $4L$ for FE NV-RAMs, where L is the *design rule* feature size), but since the flash devices work via Fowler–Nordheim tunneling, it is unlikely that they can be made to function at 1.1 V, the intended silicon logic level voltage of the

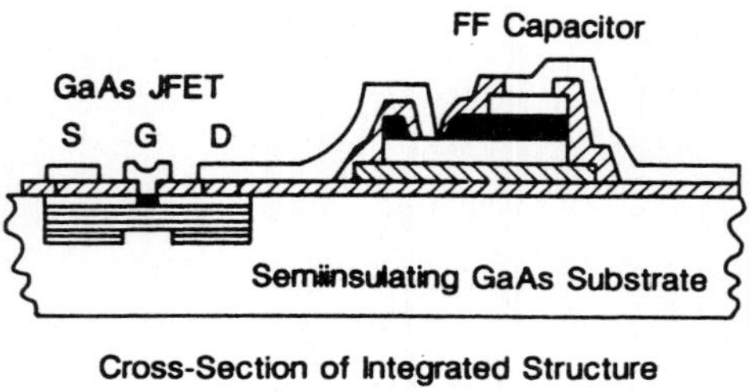

Fig. 2.35. GaAs JFET with FeRAM along side [not vertically integrated or planarized (McDonnell-Douglas)]

Fig. 2.36. 200 nm Pt post electrodes (Tegal)

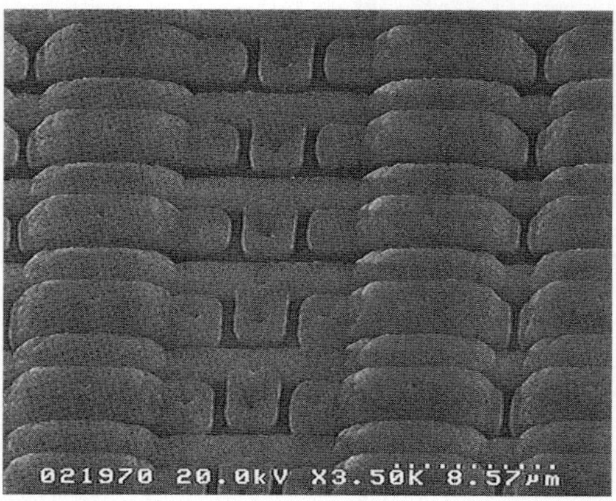

Fig. 2.37. Micron-sized ferroelectric array from National/Panasonic

(a) 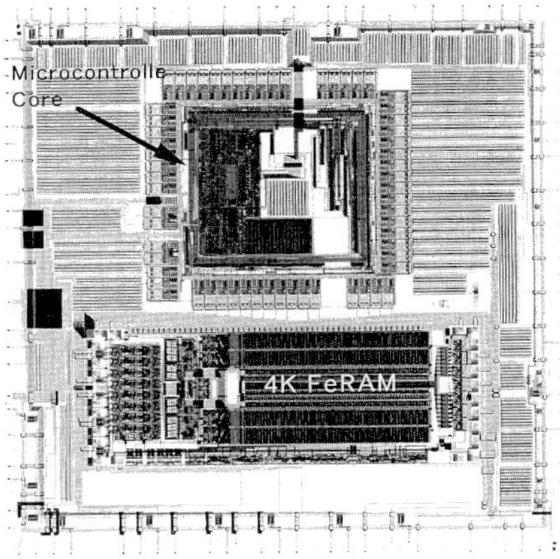

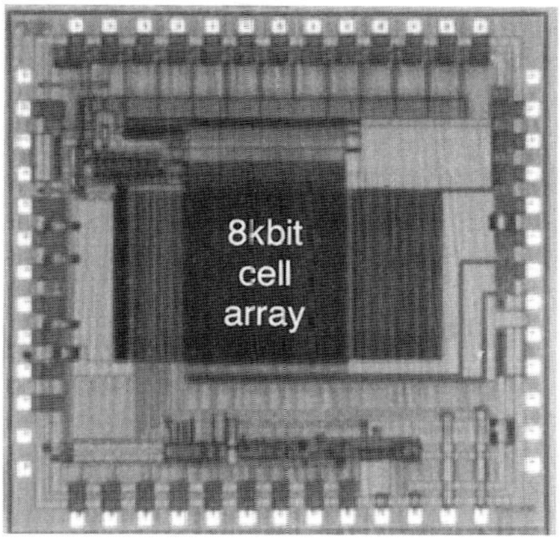 (b)

Fig. 2.38. 4K (**a**) and 8K (**b**) embedded FeRAMs

near future (it is too difficult to make reliably tunneling barriers sufficiently narrow that 1 eV electrons tunnel through them copiously). Therefore, as the silicon logic level comes down from 5.0 V to 3.3 V and finally to 1.1 V over the next few years, ferroelectric NV-RAMs should replace flash memory (Table 2.1).

It is useful to keep certain parameters in mind: for good NV-RAM operation it is necessary that the remanent polarization be $> 10\ \mu C/cm^2$, in order to provide enough switched charge to the sense amplifiers ($> 1\ \mu C/cm^2$ for a ferroelectric FET). For a DRAM the leakage current density should not exceed 100 nA/cm^2 to 1 $\mu A/cm^2$ to avoid undue heating and power consumption [62, 63, 64, 65]. For DRAM use the dielectric constant (a function of thickness) should be at least 300 at 25 nm. The breakdown voltage must exceed 3 V/25 nm = 100 MV/m.

2.3 Testing

There have been IEEE committee meetings over the past several years to determine standards for testing ferroelectric thin films for memory applications [66]. Figure 1.1 showed the hysteresis curves obtainable for real films, complete with bias voltages, trapped charge, and other 'warts' and imperfections that are often ignored in textbooks but of paramount importance in real engineering. In addition to hysteresis testing there are a number of other standard tests for ferroelectric memory films, discussed briefly below. These include $i(t)$ current transient measurements, $J(t, V, T)$ leakage current measurements, fatigue measurements of switched charge q versus switching cycles N, imprint tests to characterize preferred polarization states, retention tests $P_r(t)$, and radiation hardness tests.

Before summarizing each of these individually, let me emphasize an overall philosophy: it is essential when the results of these tests are published that all the data are for the same specimen(s). It has often occurred, particularly with results from industrial laboratories, that someone will publish, for example, a remanent polarization of 20 μC/cm^2, a breakdown field of 3 MV/cm, and a leakage current of 2 nA/cm^2 at 3 V across 180 nm. However, in reality these parameters were measured on three different specimens which had been processed in different ways to optimize one parameter. This creates, intentionally or unintentionally, the false impression that a film meeting all commercial specifications had been made. Testing engineers should publish parameters of this sort that are all taken from the same sample.

A different question is statistics. When one measures some parameters (such as breakdown field E_{B}) there is often a large spread in values for say 100 samples. In this case one is normally not interested in the average value or the low values; it is merely assumed that these are not intrinsic but are caused by imperfect contacts, etc. The value of physical interest is the *best* value; it is assumed that this is most likely to indicate the intrinsic property of the material tested. Thus, in my opinion, it is often appropriate to quote best values and not averages. However, it is helpful to readers to state clearly what is being tabulated or graphed.

There are no universal standards for testing. One can read in the literature that a ferroelectric film withstands 10^{12} bipolar switching cycles without fatigue, only to discover later that this is at 1.0 V. Generally, films should be tested at their intended operating voltage. Until now this has been 5.0 V, and authors are therefore encouraged to make measurements at that voltage.

2.3.1 Pulse Testing

The sequence of voltage pulses used in switching tests is illustrated at the top of Fig. 2.39. There are two positive voltage pulses with rise times fast compared to their widths, followed by two negative voltage pulses. In early studies (around 1990) the width of each pulse was typically 1–10 μs; the space between each pulse, typically 10–30 μs. The height of each pulse is 5.0 V. Repetition rate for the four-pulse sequence is < 50 kHz. Thus, one cycle takes > 20 μs. For a 10^{11} cycle fatigue test, this is 2×10^6 s, or one month! More modern equipment utilizes high-speed testers (and fast rise-time pulses) that permit these pulse widths and separations to be $\ll 1$ ms, thus lowering the fatigue test time to hours. Perhaps the newest commercial test apparatus at present is the Aixact TF tester from Aachen; the Radiant RT-66 is the most popular tester in use.

Early testing often utilized triangular voltage pulses instead of square pulses). This was done because it was electronically easier (no fast rise times required). However, as shown by Duiker et al. [68], the measured response from triangular voltage pulses differs from that of square pulses, so care must be exercised in interpreting such switching data.

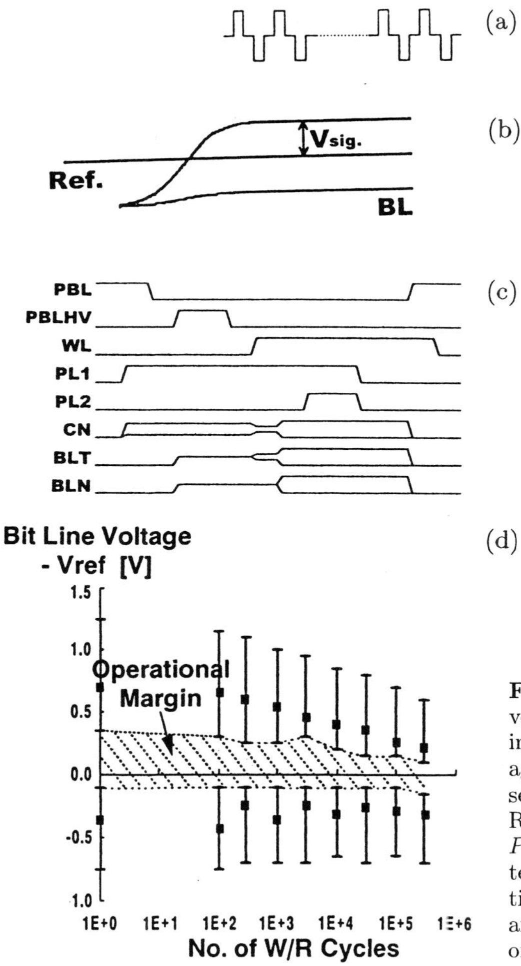

Fig. 2.39. (**a**) Sequence of four voltage pulses in a standard switching test protocol. (**b**) Signal voltage (from bit-line voltage) from a sequence of longer pulses used to READ and quantitatively measure P_r during an interruption of the test in (**a**) [58]. (**c**) Complete operational test voltage sequence (NEC) and (**d**) test results for PZT memory

It is important to include the fatigue data for the first 10^5 or 10^6 cycles, because there is often a 20% drop in this initial phase of testing. This is not due to sample heating (specimens should always be connected to a heat sink during fatigue tests), but to the time required to reach a steady state.

Recently Song et al. and Wouters et al. [69] have shown that the switching current response from the test voltage sequence in Fig. 2.39 depends strongly upon the pulse width; both coercive field E_c and remanent polarizations P_r change (E_c increases with shorter pulse width and P_r decreases). Therefore the influence of pulse width is nontrivial on the numerical values of parameters measured. This dependence upon pulse width is basically the Fourier transform of the dependence of E_c or P_r upon measuring frequency, which is discussed in greater detail in Chap. 8, but it is mentioned here because it is an important yet subtle aspect of testing protocols.

Perhaps even more important is that the apparent activation field α or activation voltage αd measured for PZT depended very strongly upon the pulse width, with αd varying from 19 V at 10 ms pulse width to 32 V at 1 μs. Stated rather more simply, this means that when we express the switching probability or switching time as activation-field dependent,

$$t_s = t_0 \exp(-\alpha/E) = t_0 \exp(-\alpha d/V) , \tag{2.3}$$

[see (6.13) Chap. 6] we should keep in mind that α is not independent of pulse width and generally decreases with increasing width.

2.3.2 $i(t)$ Current Transients

Current transient response is illustrated in Fig. 2.40. This is measured by measuring the displacement current $\mathrm{d}D/\mathrm{d}t$ through a load resistor when the ferroelectric thin-film capacitor switches polarization. In order that the measured time not be limited by RC-time constant, it is necessary that the capacitor surface area be small (usually $\ll 100\ \mu m^2$) and that load resistor be small (5 or 10 Ω). These conditions were not always met in early studies, which typically employed $A = 10^4\ \mu m^2$ capacitor areas and 1–10 kΩ load resistors. The drive voltage supply must be capable of delivering large currents with very fast rise times (around 100 ns or less). A good current transient tester is the Aixact DR tester.

It may be useful to subtract the non-switching current transient $i(t)$ from the switching transient to obtain the true displacement current. This is very important in ferroelectrics such as PZT where the large dielectric constant $\varepsilon = 1300$ makes the linear transient response comparable in magnitude to the nonlinear switching response; it is very unimportant in ferroelectrics such as $BaMgF_4$ or KNO_3 in which $\varepsilon = 20$ or so. Note that there is no correlation between the dielectric constant ε in a ferroelectric and its remanent polarization P_r; the ideal DRAM capacitor maximizes ε and has $P_r = 0$ (switching is undesirable and certainly unnecessary), whereas the ideal NV-RAM ferroelectric has large P_r and very small ε. (Thus PZT is a bad choice and $BaMgF_4$ a good choice, at least in this respect.) This last point should be emphasized in choosing materials for ferroelectric gates in FE-FETs (Field Effect Transistors), as discussed in Chap. 12.

2.3.3 Leakage Current Tests

Leakage current tests of J versus voltage at different temperatures are most useful for determining engineering heating problems (rather important for DRAM applications, but relatively unimportant for NV-RAM applications) and physical conduction mechanisms (Schottky, Poole–Frenkel,

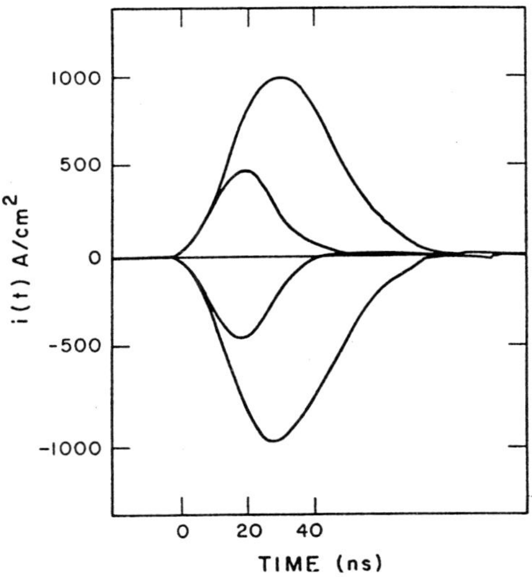

Fig. 2.40. $i(t)$ current transient from switched ferroelectric capacitor (PZT/Pt), measured as the voltage $V(t)$ across a 10 ohm load resistor

Fowler–Nordheim). As discussed in Chap. 5 and shown in Fig. 5.10, it is necessary that true leakage currents be measured in steady state. This requires waiting until a time that is long compared to relaxation times but short compared to the onset of TDDB (Time Dependent Dielectric Breakdown). This time varies with specimens and its onset is shorter in rather impure or defective samples. In pure films it may require hours at room temperature to reach the steady-state true leakage current regime. In such cases it may be necessary to make measurements at elevated temperatures.

Leakage currents should always be measured for both positive and negative voltages (the response may not be symmetrical) and for top and bottom electrodes that are of the same metal.

2.3.4 Retention Tests

Retention, as opposed to fatigue, is the ability of a ferroelectric capacitor cell to retain its stored charge q and hence its 1 or 0 of stored information. This is a shelf-life or aging test in which the cell is not cycled with bipolar applied voltages.

A retention test therefore consists of storing a bit for a time t, then addressing it to test its polarity. This is a destructive test and therefore a reset voltage must be applied after the READ operation to restore the cell to its original stored polarity. Failure of retention tests is usually due to electret effects (mobility of charged defects).

2.3.5 Imprint Tests

Testing for 'imprint' (preferred polarization state) involves a READ operation after a *waiting time* $t+$ on a cell initially polarized in a given direction (say, +), followed by a polarization reversal to the original state after a different time $t-$. Generally speaking the amount of charge switched + or − is different, and the switching times observed can be very different (factors of 2 or more). This is usually due to space charge and is discussed in some detail by Okazaki et al. [70, 71]. In fact, if the reciprocal switching time $1/t$ is plotted versus applied field E, the difference in field required to achieve exactly the same switching time for positive or negative voltages is exactly equal to the field due to the space charge in the film. For PZT this is typically 10 kV/cm, as shown in Fig. 2.41. To test this intrinsic polarity dependence one must of course use identical top and bottom electrodes. Other studies of space charge in PZT/PLZT include the LIMM (Laser Intensity Modulation Method [72]) results of Chen and Yao [73], which give the distribution of space charge $\rho(z)$ as a function of film depth.

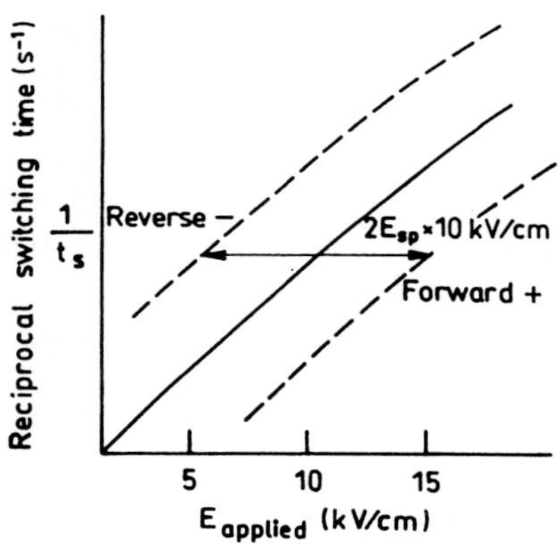

Fig. 2.41. Reciprocal switching times $1/t_s$ in ferroelectric PZT films, showing the 'imprint' asymmetry due to space charge

2.3.6 Capacitance Versus Voltage Testing: $C(V)$

In simple textbooks it is implied that measuring the capacitance $C(V)$ versus voltage for a ferroelectric is redundant to measuring the $P(E)$ hysteresis shown in Fig. 1.1. Since

$$q = C(V)V \; , \tag{2.4}$$

measuring charge versus voltage $q(V)$ in a hysteresis loop should give the same information as does $C(V)$. Unfortunately, although this is a reasonable statement for bulk ferroelectrics, it is quite untrue for thin films and ignores the electrode interface properties. This is discussed at length, quite correctly, by Brennan [74, 75]. The simple result is that $C(V)$ should always be measured and compared with $P(E)$ hysteresis data. In general the two are complementary techniques and together can be used to separate out surface/interface contributions [76]. $C(V)$ is usually measured [77] on commercial capacitance–voltage meters, such as the Hewlett Packard 4280A. There is a small ac modulation signal at around 1 MHz applied on top of a very slow (quasi-dc) sweep voltage with a sweep rate of order 1 V/s. Typical BST data are shown in Fig. 2.42.

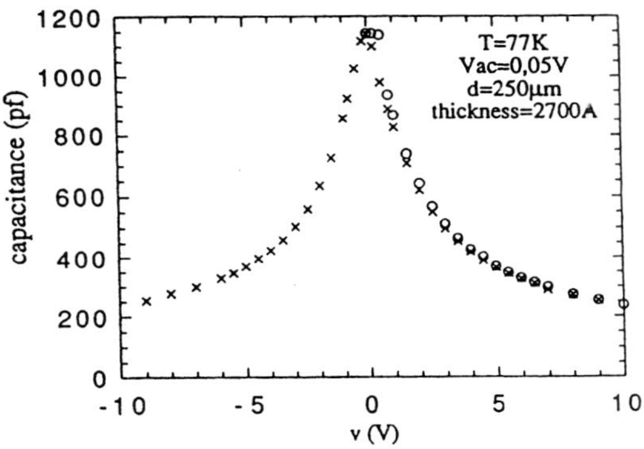

Fig. 2.42. $C(V)$ for BST at 77 K

The presence of ionic space charge layers is known in the more general case of materials with Pt electrodes to give nonlinear $C(V)$ capacitance–voltage data [78]. Both cations and anions can become chemisorbed on the Pt surfaces at voltages above 0.5 V. This is a kind of electret effect and can produce spurious hysteresis in paraelectrics, such as strontium titanate. It can be misinterpreted as ferroelectricity in amorphous materials [79].

3. Electrical Breakdown (DRAMs and NV-RAMs)

Many aspects of the design engineering, materials processing and selection, and applied physics (e.g., switching kinetics) are unrelated in ferroelectric applications to nonvolatile RAMs compared with DRAMs. In the ferroelectric NV-RAM the ferroelectric polarization contains the stored information whereas in a ferroelectric DRAM, the ferroelectric film is merely a high-dielectric capacitor and can have $P_r = 0$. However, some issues are the same, and this chapter deals with one of them, electrical breakdown, which is paramount for NV-RAMs, DRAMs, and other integrated ferroelectric devices such as bypass capacitors, which are not memories.

Figure 3.1 shows the first successful ferroelectric thin-film capacitors fully integrated into a semiconductor IC. These five BST/GaAs MMIC parts, a mixer IC (GN2012), a wideband amplifier (GN1016), a low-noise amplifier (GN1025), an integrated mixer (GN2014), and a one-chip front-end IC (GN1023) were developed in 1990–91 by Matsushita–Symetrix and in 1994 were awarded the Japanese electronics industry prize as product of the year by Nikkei Shimbun. They are used in digital mobile telephones and in TVs for the 800 MHz–2.3 GHz regime. Figures 3.2 and 3.3 show the integration and business plans for the same BST material into a 64 Mb DRAM, and Fig. 3.4 shows an SEM cross-sectional view of the BST or $SrTiO_3$ on the Pr/TiN electrodes.

The step-by-step integration of this material in a full 16 Mb DRAM by Panasonic is shown in Fig. 3.5, steps 1 to 5. The first step is the plug formation, via deposition and etching of doped polysilicon. The second step is the formation of the storage nodes. These storage nodes can be made from titanium silicide $TiSi_x$, from titanium nitride TiN on Pt, or combinations thereof. The node electrode is etched via Reactive Ion Etch (RIE). The third step is the formation of the ferroelectric capacitor and the cell plate. This uses a top electrode of TiN or Pt, and the etching is done via ion beam etching, unlike the preceding RIE step. The fourth step is planarization. This involves a flowing boron-silicate glass (BPSG) at 900°C under a nitrogen atmosphere. Then the first metal layer is deposited. This step can involve yield problems due to plasma damage to the underlying circuitry. Finally, the second metal and passivation steps are carried out. This can entail further plasma dam-

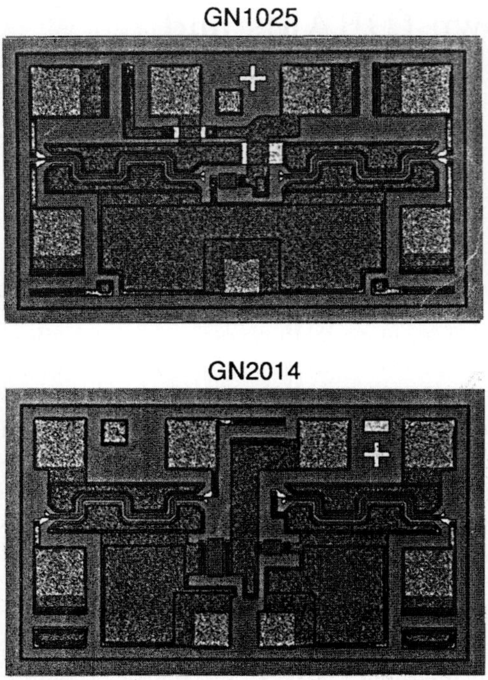

Fig. 3.1. Integrated BST/GaAs bypass capacitors (Panasonic)

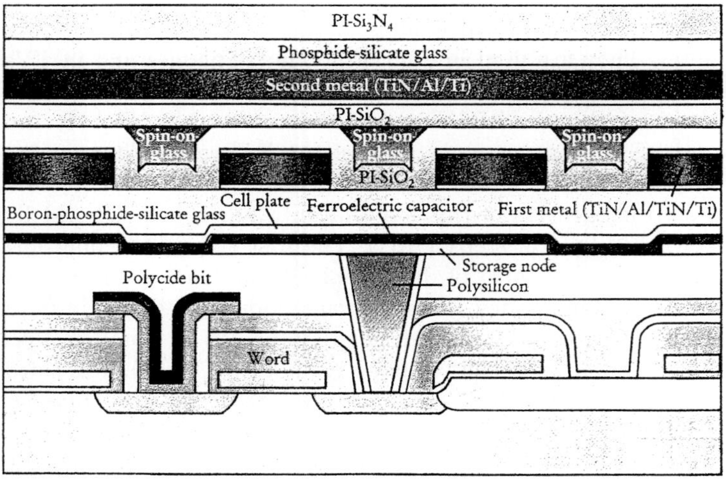

Fig. 3.2. DRAM with ferroelectric capacitor (Panasonic)

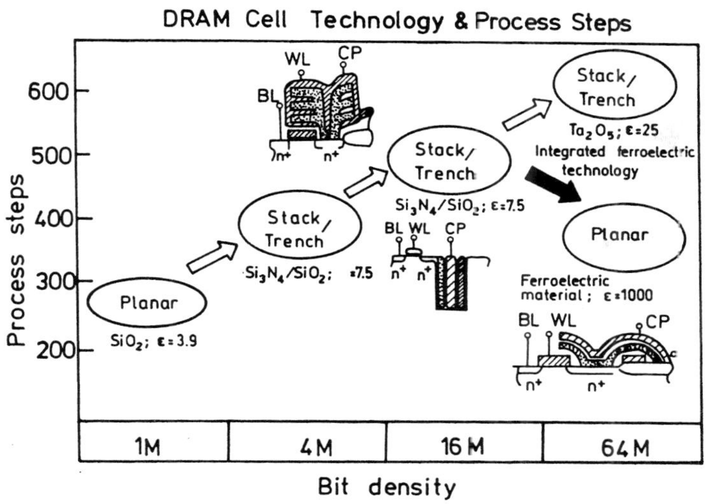

Fig. 3.3. DRAM business plans

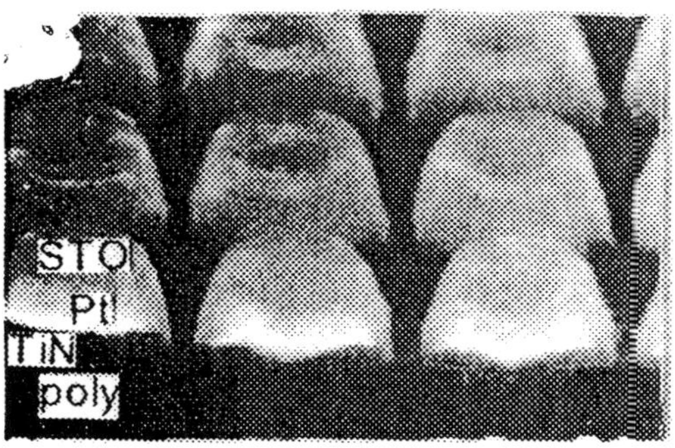

Fig. 3.4. 0.5 micron $SrTiO_3$ DRAM capacitors (NEC)

age and also the possibility of hydrogen (forming gas) contamination of the ferroelectric.

The same kind of processing steps are used for the nonvolatile RAM in Fig. 3.5. However, in this case the cell plate must be different. The NV-RAM cell uses a divided cell-plate design. This is because the cell plate has a very large area A compared with that in a conventional RAM interconnect, because it is also one of the two capacitor electrodes. This area is minimized by using a divided cell plate that reduces parasitic capacitance from 35 pF to 8 pF, which in turn reduces current consumption to 0.4 mA at 3 V and 200 ns cycle time, at a cost of increasing A a negligible 3%.

Maximum speed in a ferroelectric RAM cell is achieved by maximizing voltage at a given operating voltage. This implies that the film thickness d should be as thin as possible without suffering electrical breakdown (shorting)

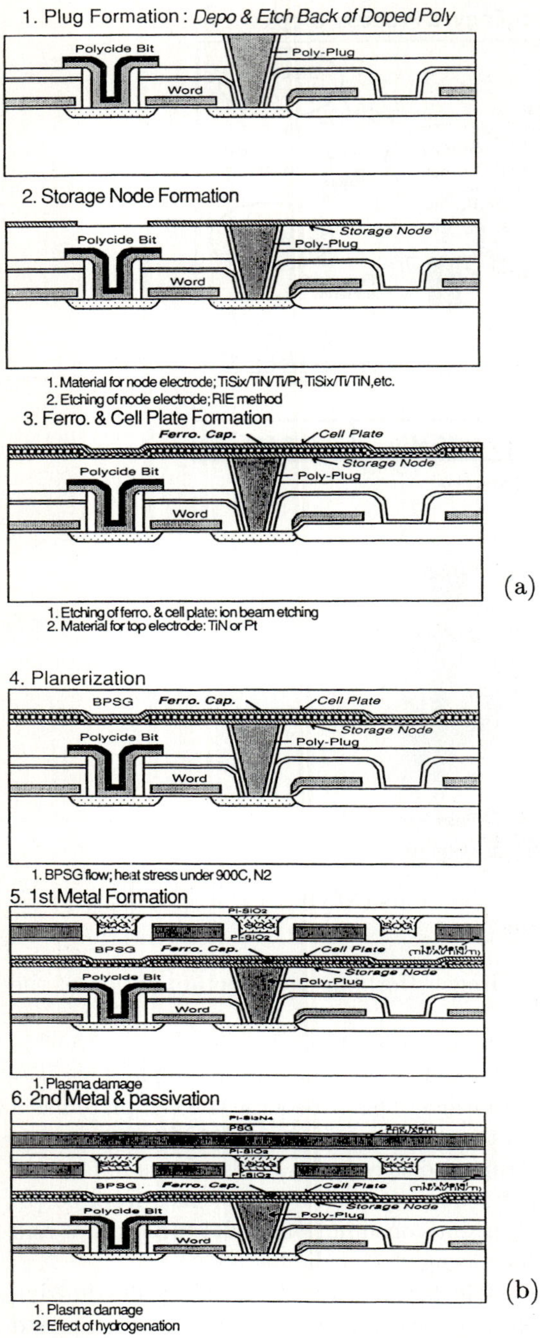

Fig. 3.5. Processing steps for a BST DRAM (Panasonic). (**a**) Steps 1–4. (**b**) Steps 5 and 6

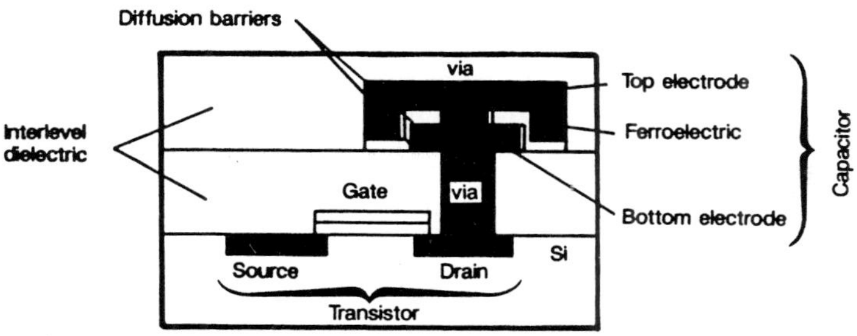

Fig. 3.6. High density 1T–1C NV-RAM

or large tunneling currents (which cause undesirable heating). In addition, the capacitance of the cell will vary as $1/d$. Thus, both capacitance and speed require that film thicknesses be minimum without risking breakdown.

Electrical breakdown is as important in DRAMs as in nonvolatile RAMs, and the physics is very much the same. In the DRAMs the oxide film must have a high dielectric constant, but the remanent polarization can be small or zero (paraelectric). BST is the most popular material for ferroelectric DRAMs, and a typical cross section is shown in Fig. 3.2.

The mechanism of dielectric breakdown in thin oxide films, including the ferroelectrics most commonly employed for memory devices, is complicated. In general, purely electronic processes (e.g., Zener breakdown) in which breakdown arises from excess electron concentrations in the conduction band do not occur. The latter require cryogenic temperatures in ferroelectric oxides and are important only for materials in which the thermal conductivity greatly exceeds the electrical conductivity (such as mica). Ferroelectric films appear to exhibit avalanche breakdown under most normal operating conditions. Avalanche is itself a hybrid process, best exemplified by spark discharge in air, which is electronically initiated but characterized over most of its discharge time by thermal runaway. Some authors use the term *avalanche breakdown* to refer only to processes initiated by impact ionization across the bandgap. However, this is very unlikely in wide bandgap ferroelectric oxides. If impact ionization occurs there, it is apt to be ionization of carriers from filled traps, not across the bandgap. For reasons discussed in Chap. 18, this avalanche breakdown is more likely to be initiated by field emission from the electrodes and not by impact ionization. The former is well known to occur in II–VI semiconductors [80] between 1.0 and 2.0 MV/cm, exactly the value found most frequently for ABO_3 ferroelectrics. Waser and Smyth have argued that breakdown in ferroelectric thin films is not avalanche, but rather impulse thermal, and point out that the more effective heat-sinking of thin films (compared with bulk) should militate against avalanche mechanisms [80]. They refer to avalanche breakdown as 'rare' but mention other electronically initiated model mechanisms. However, the view in the present book is that although impulse thermal describes many of the breakdown details, the initiation is electronic and governed by the electrode interface, contact potentials, and the Schottky barrier height. The best evidence for this in PZT or BST is that there is a very large variance in breakdown times but a very small variance in breakdown fields, compatible with avalanche. The best test, yet unmade, would be breakdown field versus thickness for very thin films; avalanche has a thickness threshold, as shown in great detail for ZnS films.

3.1 Thermal Breakdown Mechanisms

The general expression for breakdown field E_{B} is given by

$$C_{\mathrm{V}}\frac{\mathrm{d}T}{\mathrm{d}t} - \nabla \cdot (K\nabla T) = \sigma E_{\mathrm{B}}^2 , \tag{3.1}$$

where C_{V} is the specific heat, T temperature, t time, K thermal conductivity (generally a tensor), σ electrical conductivity, and E_{B} the applied electric field required to produce breakdown (electrical shorting). The two terms on the left hand side of this equation are well known in the simpler theory of melting in metals [80]: the time derivative term refers to what is termed *impulse thermal* breakdown. Over a reasonable range of ramp rates $\mathrm{d}E/\mathrm{d}t$, it can dominate E_{B}. But of course the breakdown field does not increase without limit as $\mathrm{d}T/\mathrm{d}t$ increases. At very high ramp rates E_{B} will saturate at a constant value for other reasons. This term is rigorously independent of thickness d and lateral area A. The gradient term refers to what is called *dc thermal* breakdown. This is a steady state heating process that depends on the geometry of the specimen, including both its thickness and lateral dimensions.

In general this equation must be solved numerically. However, in the approximations in which impulse thermal breakdown or dc thermal breakdown is dominant, some exact analytic results are possible. We shall see that electrical breakdown in ferroelectric oxide thin films is primarily avalanche in certain characteristics, while other qualities are accurately described in the impulse thermal approximation and in the non-ohmic (exponential) conductivity regime.

In a dielectric Ohm's law is satisfied for hopping-type conduction only if

$$aeE \ll k_{\mathrm{B}}T , \tag{3.2}$$

where a is the distance between equivalent ion sites (say, oxygen ions), e the electron charge, and k_{B} Boltzmann's constant. In the case of perovskite oxides, the parameter a is typically 0.3 nm and hence for $E > 30$ MV/m the conduction is exponential in applied electric field E and not ohmic:

$$\sigma(T) = \sigma_0 \exp\left(-\frac{\beta}{k_{\mathrm{B}}T}\right) , \tag{3.3}$$

where β is an activation energy.

In this exponential conductivity regime (3.1) can be solved exactly if the dc term is neglected:

$$E_\mathrm{B}(T,\, t_\mathrm{c}) = \left[\frac{3C_\mathrm{V}K}{\sigma_0\beta t_\mathrm{c}}\right]^{1/2} T \exp\frac{\beta}{2k_\mathrm{B}T}\,, \tag{3.4}$$

where t_c is the rise time of applied electric field in the impulse thermal regime and β is measured independently from (3.3).

We can test (3.4) in three ways: overall numerical value of E_B, dependence upon T, and dependence upon rise time t_c. For rise time t_c of about 10 ns in PZT or BST the known film parameters yield a theoretical value of about 8 MV/cm (800 MV/m). This is in good agreement with the experiments to date which give for BST 580 MV/m [81] and for SBT about 380 MV/m. (Parenthetically we note that for rather less sophisticated industrial applications, PZT on steel substrates still gives a respectable breakdown field of 50 MV/m [81].) Figure 3.7a,b shows that the dependence of E_B upon t_c is compatible with (3.4) in both BST and PZT (although a linear dependence would fit as well), but for BST only from 120–180 MV/(m·µs). At lower ramp rates (longer t_c) the breakdown field saturates at about 100 MV/m. This is due to the fact that the dc-thermal gradient term in (3.1) becomes important for such slow ramp rates. Ramp rates above 200 MV/(m·µs) were not tried. The dependence upon temperature T is shown in Fig. 3.8 and shows a decrease with increasing T [presumably dominated by the exponential term in (3.4)]. This decrease in E_B with increasing T deserves some discussion. As discussed further below, the dominant breakdown characteristic in ferroelectric thin films is avalanche. However, in Si or Ge device technology, particularly pn-junctions, avalanche mechanisms always reveal themselves with a qualitatively clear signature: breakdown due to avalanche has E_B increasing with increasing T. We show below why this is true and why it does not occur for avalanche breakdown in ferroelectric oxides.

In a material such as silicon crystals the electron mean free path $\lambda(T)$ for avalanche decreases with increasing temperature due to electron–phonon scattering. Because of this loss mechanism it requires a higher voltage at high temperatures to initiate breakdown than at low T. The formula is

$$\lambda = \lambda_0 \tanh\frac{E_\mathrm{g}}{2k_\mathrm{B}T}\,. \tag{3.5}$$

This is a rather small effect. Typically it is only 10% or so between 300 K and 500 K at $n = 10^{16}$ cm^{-3} donor concentration, and less for higher carrier concentration. In ferroelectric ceramic films this effect is negligible: firstly, because the donor concentration is around 10^{20} cm^{-3} and secondly, because mean free path is limited by T-independent scattering processes, involving grain boundaries, domain walls, and defects. Thus, the E_B varies inversely with T, unlike Si, and for different reasons, discussed below, despite the fact that it is avalanche-like.

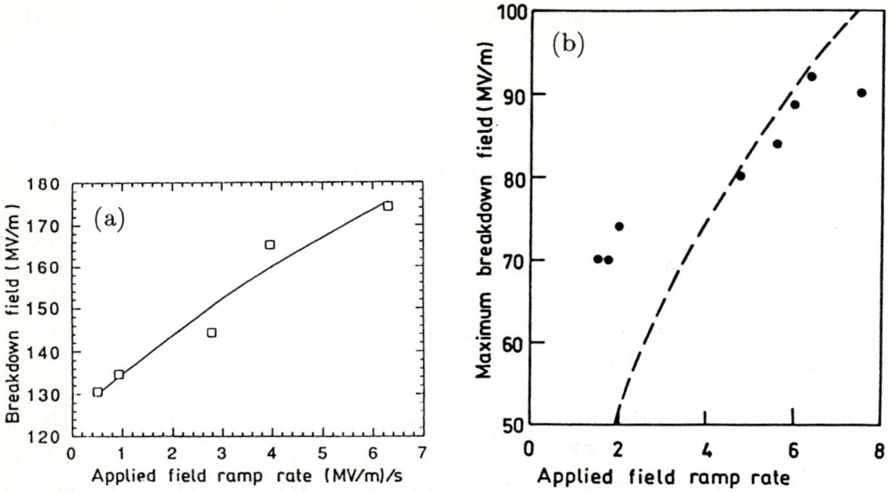

Fig. 3.7. (**a**) Dependence of breakdown field in BST/Pt upon ramp rate [92]; (**b**) for PZT/Pt [85]. Units are MV/(m·μs)

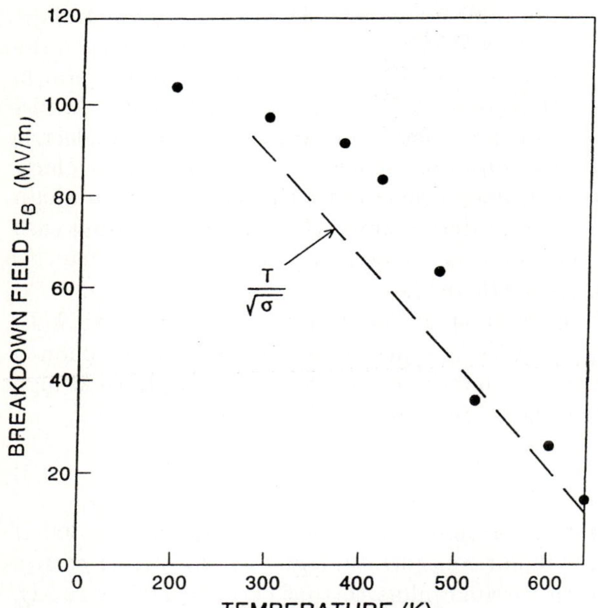

Fig. 3.8. Dependence of breakdown field in BST/Pt upon temperature [92]

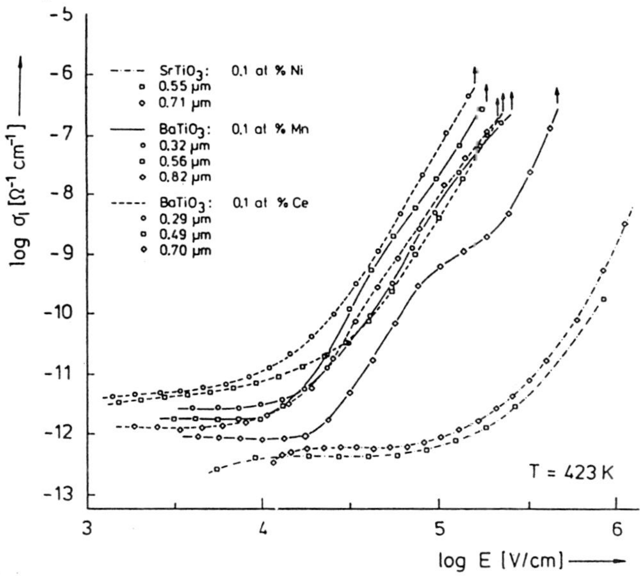

Fig. 3.9. Leakage current versus field in doped barium and strontium titanate, showing failure at a constant current density, not field [85]

Avalanche breakdown in other wide bandgap semiconductors such as ZnS is well known [82, 83, 84] and easily proved by monitoring the dependence of electroluminescence as a function of film thickness d. Typically for $d < 100$ nm the intensity of the luminescence drops rapidly with decreasing thickness because d is too thin for avalanche to develop (an exponential process).

The fact that the final stage of electrical breakdown in ferroelectric thin films is caused by thermal runaway is shown clearly in Fig. 3.9. In this figure [85] it can be seen that independent of field or thickness, breakdown occurs at a fixed current density level requisite to achieve localized melting temperatures.

3.2 Von Hippel Equations

Although we have described breakdown in ferroelectric thin films above as basically thermal, the initiation stage is not thermal but electronic (like spark discharge in air). An early theory relating the initiation of breakdown from a metal surface into a dielectric medium (gas, liquid, or solid) to electronic properties was developed by Von Hippel in three papers [86, 87, 88]. He wrote

$$eE_{\mathrm{B}}\lambda = ch\nu \ , \tag{3.6}$$

where e is the electron charge and c, a constant of order unity. The energy $h\nu$ is related to the metal work function Φ. λ is the mean free path of an electron injected from a metal cathode into the dielectric and at low voltages it will be not a constant but linear in voltage V [89]. This model was successful in giving the correct relative breakdown fields for iron and aluminium electrodes on gases [90] but gave qualitatively incorrect results for metal electrodes on liquid dielectrics [91]. It was extended to metal electrodes on ferroelectrics [92] and rewritten as

$$eE_{\mathrm{B}}\lambda = c(\Phi_{\mathrm{M}} - \Phi_{\mathrm{FE}}) , \tag{3.7}$$

where Φ_{M} is the work function of the metal electrode and Φ_{FE} is the work function of the ferroelectric semiconductor, which is its electron affinity χ plus a trap energy Δ (Δ is typically about 0.9 eV in BST, PZT, or SBT). Figure 3.10 shows that this modified Von Hippel–Scott formula fits the breakdown field data for PZT on six different elemental metal electrodes very well with $c = 1$ assumed. The calculated mean free path λ from these data is about 0.1 nm at room temperature in the PZT dielectric. (Note that, because this mean free path is less than the ferroelectric film thickness, the usual Schottky equation is invalid.) Similar data for BST are shown in Fig. 3.11 but for only three different metals, rather than six. And Fig. 3.11 also illustrates the important point that the current carriers are only electrons injected from the cathode (despite the fact that both PZT and BST are p-type in bulk). This is made clear by the fact that changing the cathode metal (solid circles) changes the breakdown field by about 50%, whereas changing the anode material (open squares) does not change the breakdown field at all. Note that, as shown in Fig. 3.12, a single monolayer of oxygen on Pt electrode surfaces raises its work function by 0.3 eV [93, 94]. Such a layer may be present in as-deposited Pt electrodes and may increase in thickness with cycling of voltage (fatigue). We return to this point later. Very recently Zafar et al. have confirmed this model by showing that the Schottky barrier height of BST/Pt decreases by 0.18 eV after prolonged bipolar voltage 'stressing' [95].

In addition to the dependences upon electrode work function, temperature and ramp rate, various models of breakdown in ferroelectrics make explicit predictions for the dependence upon capacitor area A, film thickness d, and other variables. These are discussed below.

The dependence of breakdown field for PZT/Pt upon film thickness is shown in Fig. 3.13 [96], together with the theoretical predictions from dc thermal, impulse thermal, avalanche, and a linked extended-defect model [97]. The data best fit the dependence predicted from avalanche mechanisms. Note that dc thermal models result in absolutely no dependence upon geometry, including film thickness and are therefore easily ruled out.

Figure 3.14 illustrates the dependence of breakdown field upon capacitor lateral area A for strontium titanate, PZT, and SBT. The data for $SrTiO_3$ are particularly revealing in that they cover two decades from 10^2 to 10^4 μm^2 [98, 99, 100, 101]. Figure 3.15 shows similar data for SBT [102] over nearly three decades of area. In each case it is observed that E_{B} decreases logarithmically with A. This dependence is derived below.

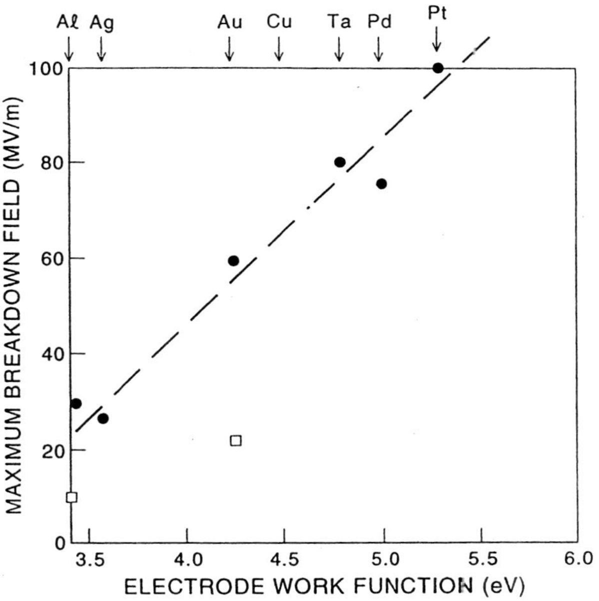

Fig. 3.10. Dependence of breakdown field upon cathode work function for PZT [96], showing agreement with (3.7). *Solid circles* show different cathode materials with the same Pt anode. *Open squares* show different anode materials with the same Pt cathode

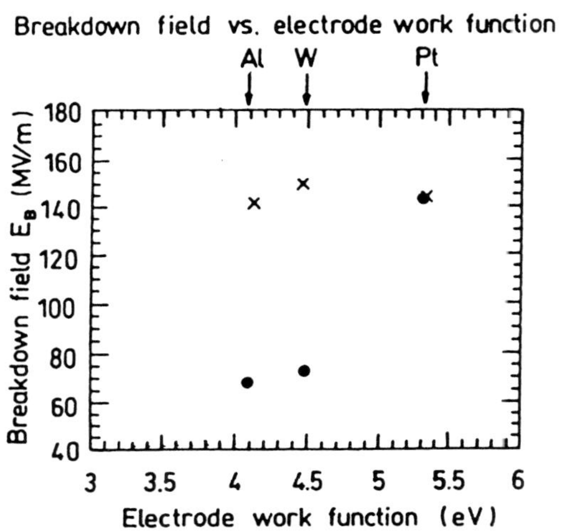

Fig. 3.11. Dependence of breakdown field upon cathode work function in BST [92]. *Solid circles* show different cathode materials with the same Pt anode. *Crosses* show different anode materials with the same Pt cathode

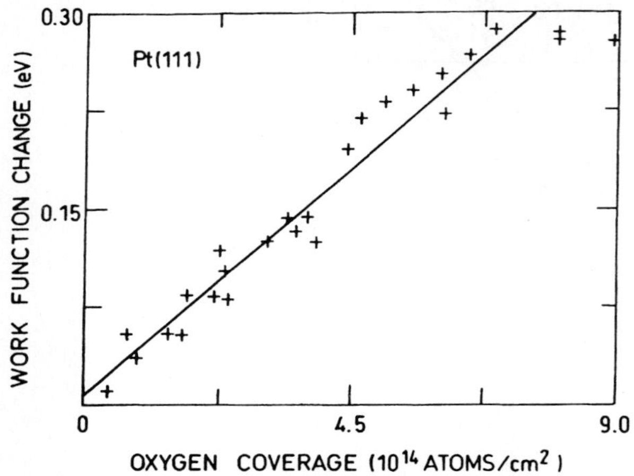

Fig. 3.12. Change in work function Φ_M for Pt [111] electrodes as a function of sub-monolayer oxygen coverage [93]

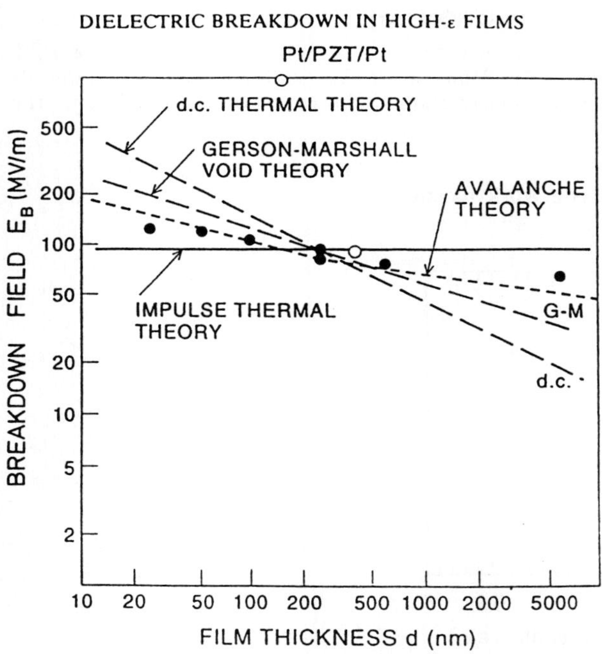

Fig. 3.13. Breakdown field versus film thickness for PZT on Pt electrodes [96]

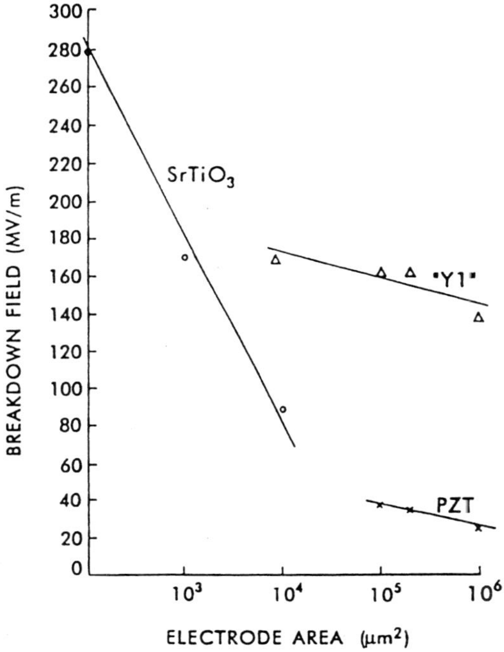

Fig. 3.14. Breakdown field versus capacitor area A for strontium titanate and PZT [98, 99, 100, 101]

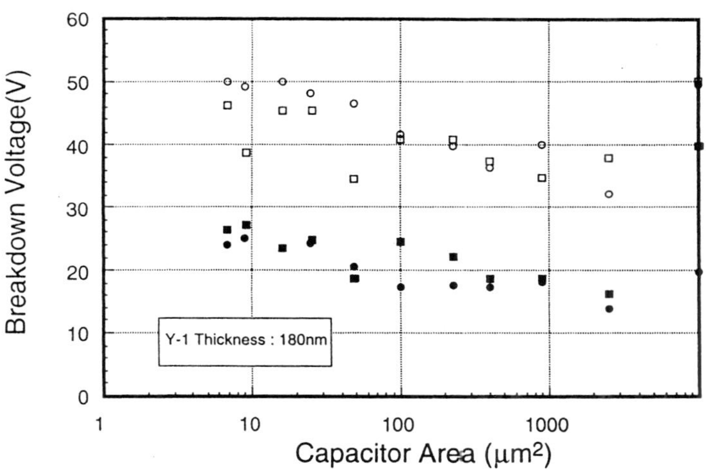

Fig. 3.15. Breakdown field versus capacitor area A for SBT [102]

If we assume that a primary conduction mechanism in ferroelectric oxide films involves hopping of electrons from site to site, and that defect and/or vacancy sites play a key role in this process, then we can calculate both the time dependence of breakdown and the areal dependence of breakdown field.

We begin with the assumption that there exists a Maxwell–Boltzmann distribution of defect energies that can produce micro-shorts. At higher fields more defect shorting paths become possible:

$$N(E, T) = N_0 \exp \frac{BE}{k_B T} , \tag{3.8}$$

where N is the number of micro-shorts and BE is an activation energy. B has units of effective charge and is assumed to vary slowly or not at all with temperature and field.

Assuming an isotropic distribution of such short-producing pathways from the electrode centre,

$$N(r) = r^2 N(E, T) , \tag{3.9}$$

we find that the probability of producing one short at a field E and temperature T is

$$1 = N_0 r^2 \exp \frac{BE}{k_B T} , \tag{3.10}$$

from which

$$\frac{BE}{k_B T} = -\log N_0 - \log A + \log \pi , \tag{3.11}$$

where $A = \pi r^2$ is the electrode area. Hence

$$E_B(T, A) = \text{constant} - \frac{k_B T}{B} \log A . \tag{3.12}$$

This predicts that breakdown field decreases linearly with increasing T (Fig. 3.8) and logarithmically in capacitor area A (Figs. 3.14 and 3.15).

The breakdown time also has a logarithmic dependence [94]. This follows more simply from the assumption of exponential conduction (3.3). If current density J is given by

$$J = \frac{q}{t} = \sigma_0 \exp \frac{\beta E}{k_B T} , \tag{3.13}$$

then

$$\frac{q}{t_B} = \sigma_0 \exp \frac{\beta E_B}{k_B T} , \tag{3.14}$$

where t_B is the breakdown time. Hence,

$$\log t_B = g_0 - g_1 E_B , \tag{3.15}$$

where g_0 and g_1 are constants. That is, the log of the breakdown time is equal to a constant minus a term proportional to the field at which breakdown

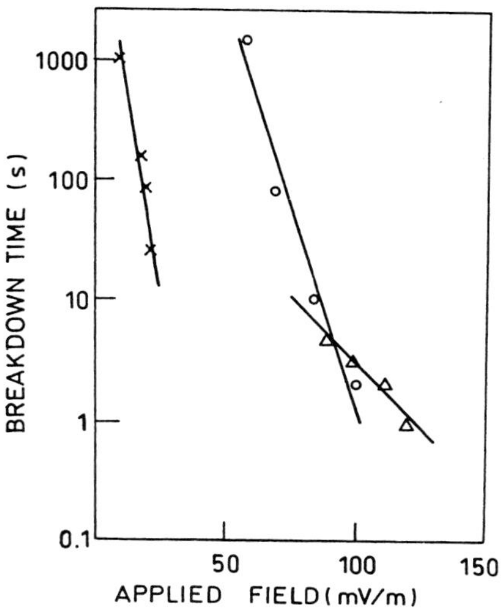

Fig. 3.16. Breakdown time t_B versus field in BST (*crosses*), PZT (*circles*) and SBT (*triangles*) [104]

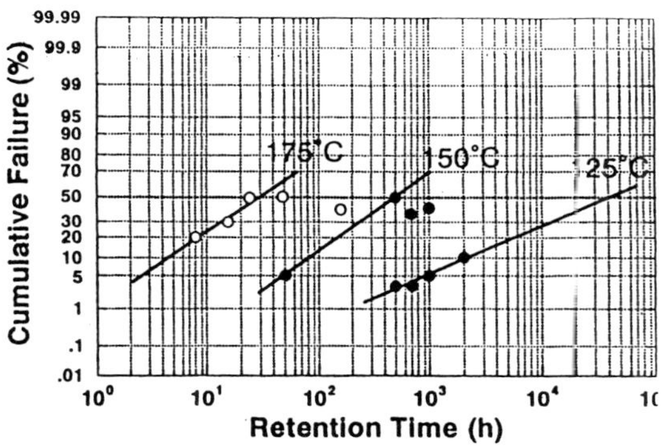

Fig. 3.17. Cumulative failures versus time in BST at very high voltages (20, 21, and 22 V across 220 nm) [92]. Note that this is around 20 times the operating voltages for 1.1 V DRAMs

occurs. This predicted dependence is confirmed in Fig. 3.15 for BST [94]. Figure 3.17 shows cumulative failure results for $d = 220$ nm BST on Pt, and Fig. 3.18 compares retention at 343 K for SBT and PZT.

Although the theory outlined above is sufficient to estimate the mean values for breakdown times, it tells us nothing about the variations and statistical distributions of these times. The latter are shown for BST in Fig. 3.19 for a 144 nm thick film [92]. Here the breakdown times vary by about two orders of magnitude. Such a large variation in breakdown time, together with a very small variation in breakdown fields, is compatible only with avalanche mechanisms.

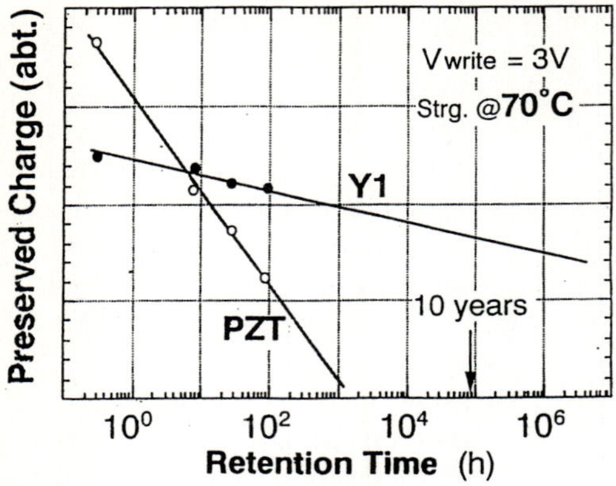

Fig. 3.18. 70°C retention in SBT and PZT

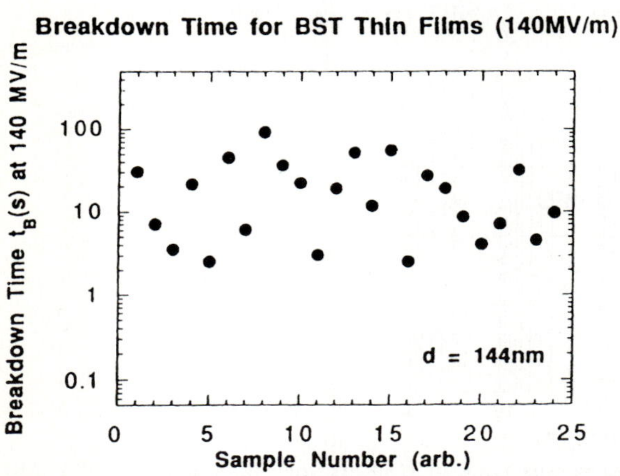

Fig. 3.19. Variation in breakdown time t_B at a constant field of 140 MV/m across a 144 nm thick BST film (Pt electrodes). The data are in accord with avalanche models [104]

3.3 Dendritic Breakdown

The mechanisms for breakdown discussed above are not very specific with respect to microscopic structure. But in fact many studies, both theoretical and experimental, show that breakdown consists of microscopic electric shorts that have finger-like dendritic structure reaching out from cathode and/or anode into the interior of the films in ferroelectrics. In general these dendritic shorts will start at inhomogeneous defects at the electrode–dielectric interfaces. Both atomically rough spots and oxygen concentration variations can initiate such dendritic shorts. Their propagation through the ferroelectric films is less well understood. However, it plays a key role in the theory of Gerson and Marshall for breakdown [97], and we have some information on extended defect structures idiosyncratic to perovskite oxides. It is possible that the linked, extended defects postulated by Gerson and Marshall for PZT play a role in the propagation of such dendrite-like tips in both ABO_3 perovskite ferroelectrics and SBT-family Aurivillius layer-structure perovskites, in which the TiO_6 octahedra are interspersed with Bi_2O_2 planes. As shown in Figs. 3.20 and 3.21 very specific kinds of defects are 'intrinsic' to the SBT family of layer-structure perovskites. These involve 'missing' TiO_6 octahedra, with consequent lattice discontinuities of 1/4 lattice constant (HREM study of $SrBi_2Ti_4O_{15}$ [104]) and result in lines of dislocation or disclination which always meet in groups of four (never three) at each vertex. These AntiPhase Boundaries (APBs) are also well known in tungsten bronzes [105] such as $Ba_2NaNb_5O_{15}$ (Fig. 3.22) and are described by a two-dimensional four-state vector Potts model (*clock model*) in Fig. 3.23 [106]. Here the four states represent tilting of the TiO_6 octahedra along the four nominally equivalent a-axes perpendicular to the unique tetragonal c-axis. Note that a four-state scalar Potts model fails to give this result and instead predicts threefold vertices as the stable defect structure. Perhaps the most important thing to point out about APBs is that they are very different from domain walls. This is most easily visualized in a simple perovskite structure such as $SrTiO_3$. Below 105 K strontium titanate becomes tetragonal, and there are rectilinear tetragonal domain walls. In addition, however, there are antiphase boundaries across which the sense of rotation of the TiO_6 octahedra changes from clockwise to counter clockwise. These APBs are not rectilinear; they are rather wiggly. Both clockwise and counterclockwise phases have the same tetragonal axis; that is, the APBs lie within a single tetragonal domain. The behaviour of domain walls and APBs at low temperatures in $SrTiO_3$ are quite different.

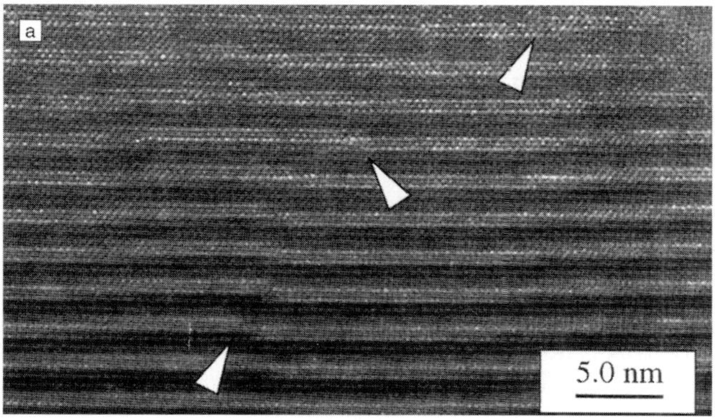

Fig. 3.20. High resolution electron microscopy of $SrBi_4Ti_4O_{15}$, showing $a/4$ defect structure [105]

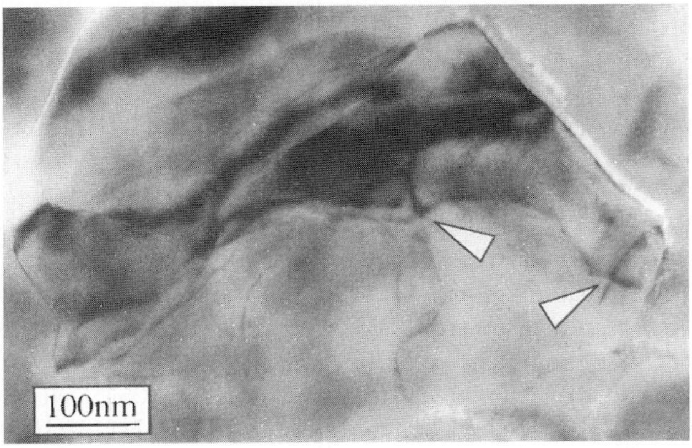

Fig. 3.21. High resolution electron microscopy of $SrBi_4Ti_4O_{15}$ showing fourfold vertices of $a/4$ antiphase boundaries [105]

(a)

(b)

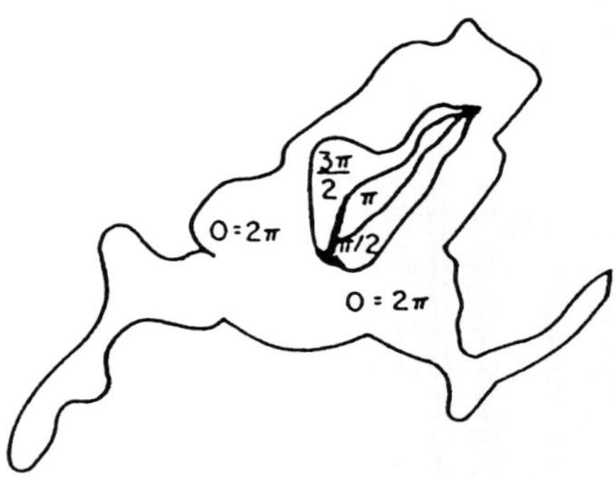

Fig. 3.22. (**a**) Fourfold vertices of antiphase boundaries in $Ba_2NaNb_5O_{15}$. (**b**) Dynamics of threefold vertices (unstable) and fourfold vertices (stable) in $SrBi_4Ti_4O_{15}$ or $Sr_2Bi_4Ti_5O_{18}$ [105]

In the early study of electrical breakdown in PZT bulk ceramics [107] it was observed visually that shorting was accompanied by the growth of dark filaments through the ceramic sample, visible through low-power optical microscopy. These are interpreted as growing 'fingers' of highly conducting regions of space charge. Although Plumlee did not illustrate these filaments in his Sandia report, a rather good illustration of similar 'dendrites' is shown in Fig. 3.24, where $LiNbO_3$ ferroelectrics were left under a 200 keV electron

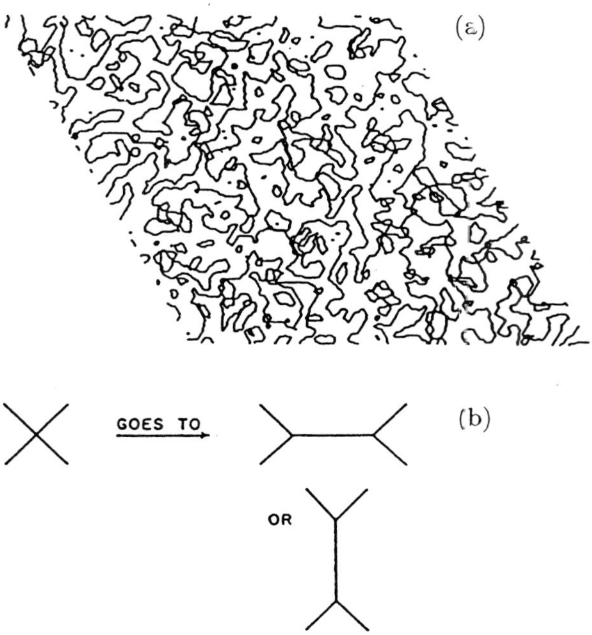

Fig. 3.23. Fourfold vertex theory for antiphase boundaries [106]. (**a**)Resulting antiphase boundaries. (**b**) Stability criteria

beam [108]. Note that the dendritic structures are long and slender (prolate) with branching. The scale is of order 1 micron, full height. A good indication that such structures develop in ferroelectric memory films undergoing repetitive voltage cycling (fatigue) is that the switching speed increases just before breakdown. This is due to the fact that as the dendritic fingers extend from anode to cathode, the voltage drop occurs from tip-to-tip, a distance d' much smaller than the film thickness d. Thus, at constant applied voltage, the local electric field increases dramatically just before shorting, which in turn increases the switching speed (roughly, as $E^{-3/2}$).

Rather different shapes for conducting shorts are observed in PZT films with oxide electrodes [109] and with Ag-paint electrodes [110]. In this case rather large (micron diameter) pancake-like (oblate) shorts result. The difference has been explained by Raleigh [111]: the growth of dendritic filaments of oxygen-deficient material in oxide ceramics is favoured with oxide electrodes; on such electrodes a concentration variation of oxygen occurs such that nucleation sites for inhomogeneous nucleation are plentiful. This should not occur for elemental electrodes such as Pt, but it can be a serious problem with IrO_2, RuO_2, or $SrRuO_3$ electrodes.

Such dendritic trees were modeled theoretically [112, 113, 114] to explain both fatigue and breakdown and also supported independently by Freund

Fig. 3.24. (*Left*) dendritic 'trees' formed by injection of high-energy electrons into an ABO_3 ferroelectric [108]. (*Right*) similar $BaTiO_3$ whiskers formed by 2 nm diameter e-beam

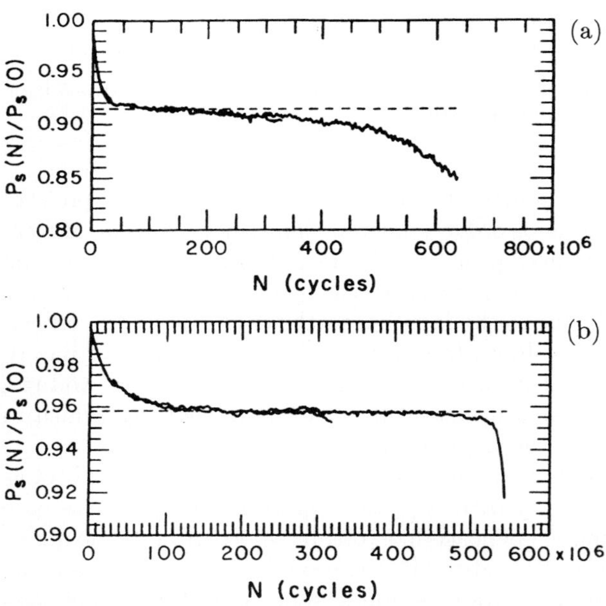

Fig. 3.25. Theoretical remanent polarization in a ferroelectric film from extended defect model [112, 113, 114]. **(a)** *Top*: soft failure. **(b)** *Bottom*: hard failure

et al. [115]. Duiker's model assumes that the 'fatigue' (apparent decrease in switched charge with repetitive bipolar voltage cycling) arises from oxygen vacancies and that when an oxygen vacancy touches an existing dendritic tree it can stick but, depending upon temperature, later hop off. However, if such a vacancy touches two sites on an existing dendritic tree, it is irreversibly attached. His results for apparent remanent polarization P_r as a function of switching cycles N_s are shown in Fig. 3.25. Here the top trace represents a 'soft' failure in which a linear proportionality between vacancy generation and applied field is assumed. The bottom trace is a 'hard' failure in which the proportionality is exponential. The latter hard failure should occur at lower temperatures, according to the Boltzmann probability of having oxygen vacancies stick to the dendritic tree. The traces each exhibit three distinct regimes. For the first 10^7 voltage cycles (represented by 10–20 time steps in the simulation) there is a rapid decrease in apparent remanent polarization; this is the time required for the oxygen vacancy concentration to reach steady state. Then there is a long flat plateau in which polarization is constant; this is the time in which dendrites are growing but remain short compared with the film thickness d. This takes typically 5×10^8 voltage cycles or roughly 500 time steps in the simulation. Finally, there is a regime of logarithmic or more abrupt decrease of polarization to zero; in this regime much of the film is screened out from participation in the switching process, as shown by the equipotential graph in Fig. 3.26 and especially the field contours in Fig. 3.27.

This model corresponds very well to the data shown for bulk PZT at room temperature in Fig. 3.28 [116] and for PZT films at various temperatures in Fig. 3.29 [117]. Note that the failure becomes 'harder' at 243 K compared with 373 K. Note also that typically published experimental omit the data for the first 10^7 cycles. Figure 3.30 shows such data and we note the short-time drop predicted theoretically is indeed observed in this regime, confirming Duiker's model predictions.

Almost all breakdown mechanisms have the qualitative result that breakdown field (not voltage) decreases with increasing film thickness d. The only exception, in which breakdown field E_B actually increases with increasing film thickness d, is called *streamer* breakdown. It is appropriate for single-point shorting on thick electrodes [118]. It predicts a dependence of

$$E_B = \text{constant } \log \frac{d}{\mu \tau E_B} \, . \tag{3.16}$$

Here μ and τ are mobility and collision time of carriers. This transcendental equation in E_B can be solved graphically or numerically to yield $E_B(d)$, and as shown in Fig. 3.31 appears to satisfy the data for some barium titanate films [119].

By comparison, collision ionization when the initiating electrons are injected via field emission from the cathode yields [120]

$$E_B = \text{constant } d^{-w} \, , \tag{3.17}$$

Fig. 3.26. Equipotential curves in a fatigued ferroelectric thin film [112, 113, 114]

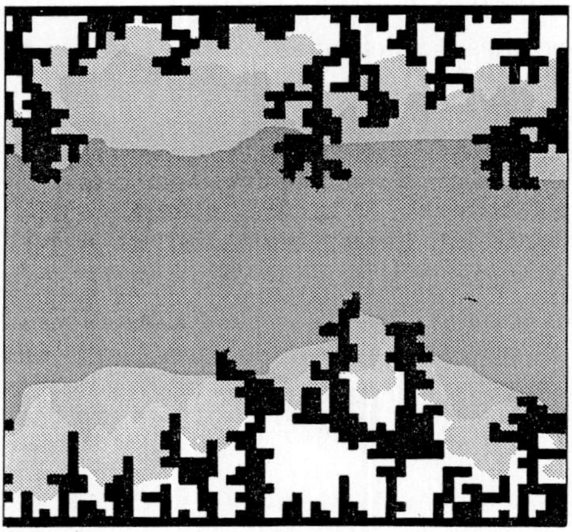

Fig. 3.27. Field contours in a fatigued ferroelectric thin film [112]

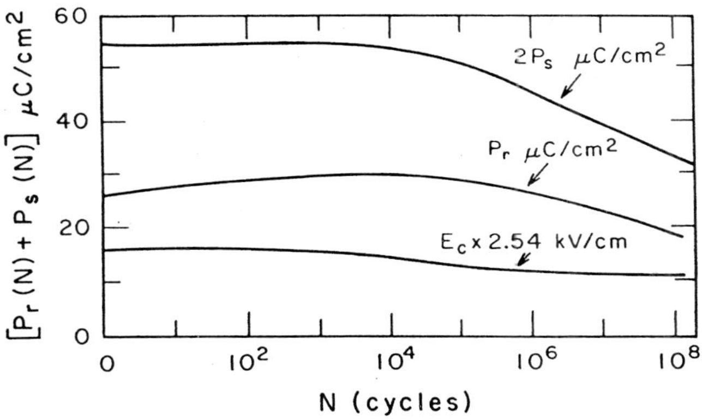

Fig. 3.28. Experimental data on decrease in remanent polarization versus switching cycles for bulk PZT [107]

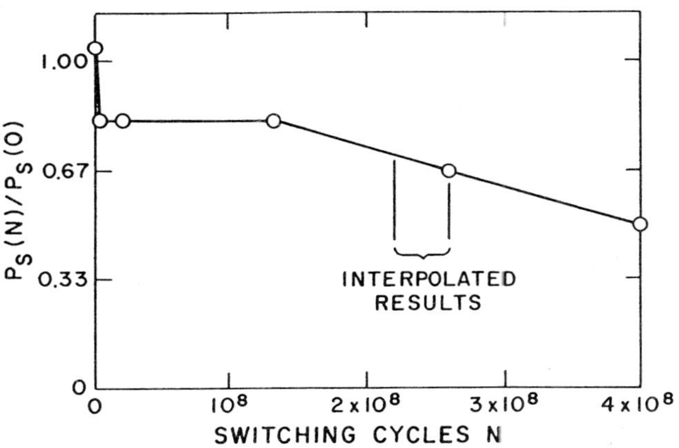

Fig. 3.29. Experimental data on decrease in remanent polarization versus switching time in PZT thin films, showing predicted decrease in the first million cycles as the oxygen vacancy concentration reaches steady state [116]

with $1/4 < w < 1/2$. Although not originally compared with experiment in Fig. 3.13, this slow dependence upon d is not incompatible with the PZT/Pt data. This kind of field emission-driven collision ionization may in fact be the best description of the initiation stage of electrical breakdown in ABO_3 ferroelectrics.

Joule heating thermal runaway satisfies [121, 122, 123, 124, 125] in the case (3.13,3.14) of exponential conduction

$$E_{\mathrm{B}} = \text{constant } \log \frac{K}{aed\sigma_0 A E_{\mathrm{B}}^2}\,, \tag{3.18}$$

[parameters defined in (3.3) and (3.4)], which also describes the experimental data in Fig. 3.13 on PZT/Pt very well.

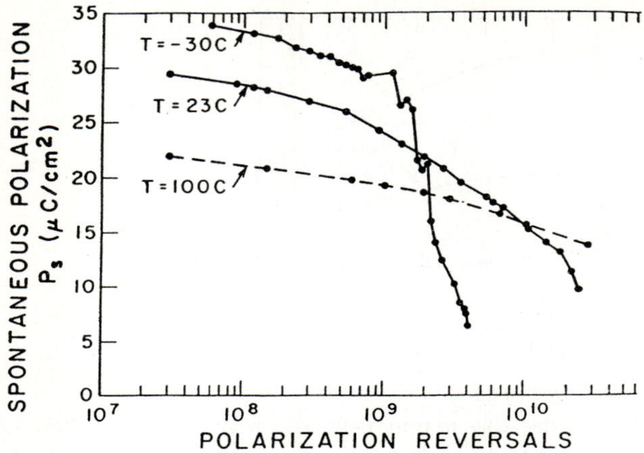

Fig. 3.30. Experimental data on decrease in remanent polarization versus switching cycles for thin film PZT [96]

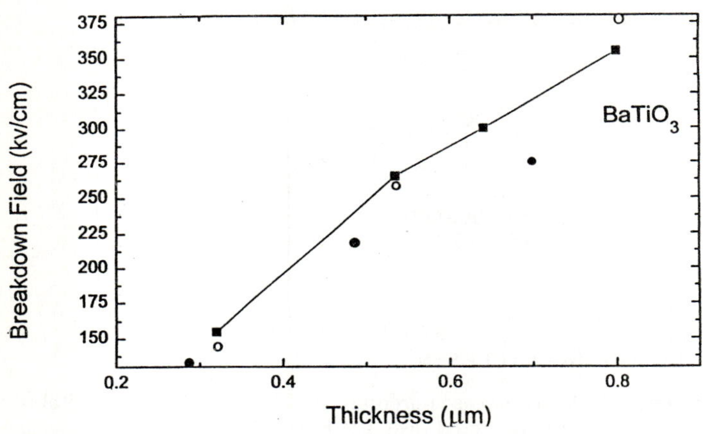

Fig. 3.31. Breakdown field versus thickness in barium titanate. Data from Waser and Klee [119]

Finally, there is an extrinsic beakdown mechanism involving the breakdown of surface layers that lie adjacent to the electrodes. These layers may be amorphous or crystalline, but in either case they have a density of surface states in the forbidden band gap. In such situations there can be interfacial breakdown which varies as

$$E_{\mathrm{B}} = \text{constant } \exp \frac{e\Delta V}{k_{\mathrm{B}}T}\,, \tag{3.19}$$

where $e\Delta V$ is the binding energy of the surface states. Published data for ferroelectric ABO_3 thin films are not in agreement with this model.

3.4 Breakdown Voltage Asymmetry and Leakage Current Asymmetry

Tagantsev et al. [126] have shown that the large asymmetry in leakage current and breakdown fields in ferroelectric thin-film capacitors, observed even with nominally identical top and bottom electrodes, results from the lowering or raising of the Schottky barrier height due to polarization. This effect was first reported by Bernacki [127]. In the model of Tagantsev et al., it is explained by simply replacing the applied electric field E in the simple Schottky equation by $E = (V/d) \pm aP_\mathrm{r}/\varepsilon_0 d$, where a is proportional to the thickness of the passive layer at the electrode–ferroelectric. Scott et al. find that a is about 20 nm in PZT [128]. Tagantsev et al. find $a/\varepsilon_0 = 0.02$ nm; that is, $a = 26$ nm, in good accord with [128]. This predicts a breakdown voltage asymmetry of approximately 2 V, in accord with recent measurements at SONY on SBT [129] as well as with the PZT results [126], which gave an asymmetry of about 100 kV/cm over 200 nm $= 10^7$ V/m $\times 2 \times 10^{-7}$ m $= 2$ V. A smaller contribution to breakdown voltage asymmetry is the thermal gradient at the electrode interface, which gives an additional ± 0.1–0.2 V through the pyroelectric effect [130].

4. Leakage Currents

It is important to understand the nature of leakage currents in ferroelectric memories in order to control heating of the memory and thermal breakdown. In general, ferroelectric film capacitors on metal electrodes provide very complicated structures in which electrons, holes, and ions all contribute to conduction, and where a variety of mechanisms – including surface-limited processes such as thermionic Schottky currents and quantum mechanical Fowler–Nordheim currents, as well as bulk-limited processes such as Poole–Frenkel and Space-Charge-Limited Currents (SCLC) are all important. Typically these processes are all present at the same time, and unfortunately they are not additive; rather, they result in nonlinear integral equations for total leakage current $J(V)$. In this book we will use the J to denote real leakage current and the letter i to designate displacement current.

4.1 Schottky Emission

Schottky barriers and the resulting $J(V)$ current–voltage relationships are discussed at length in many textbooks. A good summary is given by Sze [131] and the relationships of importance are shown diagrammatically in Fig. 4.1. Here the metal work function Φ_{M} is the voltage difference between the metal conduction band and the vacuum level (a free electron just outside the metal surface). Similarly, the difference between the ferroelectric (semiconductor) conduction band and its vacuum level is called the electron affinity χ. Note in this diagram that the vacuum level of the ferroelectric is not equal to the vacuum level of the metal, due to charging effects. Note further that the electron affinity χ is not equal to the ferroelectric work function Φ_{FE}. The potential energy difference between electron charge times the metal work function $e\Phi_{\mathrm{M}}$ and the ferroelectric work function $e\Phi_{\mathrm{FE}}$ is called the contact potential [131]. The ferroelectric work function is equal to χ plus the energy difference between the ferroelectric conduction band and the ferroelectric Fermi energy, which may be pinned by donors or acceptors. For a small barrier width w in a material which is ionic and with no surface donors or acceptors, the barrier height Φ_{B} is exactly equal to

$$\Phi_{\mathrm{B}} = \Phi_{\mathrm{M}} - \chi \,, \tag{4.1}$$

and the derivative of the barrier height with respect to electron affinity is

$$S = \frac{\mathrm{d}\Phi_\mathrm{B}}{\mathrm{d}\chi} = 1.0 \,. \tag{4.2}$$

Conversely, if a material has a large surface charge density, due either to traps (donors and acceptors) or to the covalent nature of the material, then the barrier height is pinned by these charges and is independent of the metal work function. The space charge at the interface is unaffected by changes in the metal used for the electrode:

$$S = \frac{\mathrm{d}\Phi_\mathrm{B}}{\mathrm{d}\chi} = 0 \,. \tag{4.3}$$

In real ferroelectrics the situation is intermediate between these limiting cases, and S is around 0.5.

It is clear from Fig. 4.1 that in general there will be a depletion layer inside the ferroelectric near the metal electrode. The charge density $\rho(z)$ as a function of depth z into the ferroelectric is derived from Poisson's equation. This results generally in a decrease in charge density with z such that the half-width of $\rho(z)$ is in this book termed d_d, the depletion width. If this width is small (around 2–7 nm), Fowler–Nordheim currents (quantum mechanical tunneling) can flow through the barrier. This typically occurs in ferroelectric oxides at fields above 40 MV/m, which is only 8 V across 200 nm. Some authors have argued that the depletion width d_d in thin films of PZT, BST, or SBT exceeds the film thickness d (around 200 nm). We believe that this is not true, and that depletion widths in ferroelectric films are typically 5 nm at $V = 0$. This is easily proved by the observation of Fowler–Nordheim tunneling in such films [132, 133, 134]. At smaller fields, Schottky emission is dominant.

The Schottky effect is the lowering, due to image-field corrections, of the energy required for electrons to go over the barrier at an interface when an electric field is applied. At a finite temperature electrons can thermally hop over the potential barrier shown in Fig. 4.1 even at very small applied voltages. These transitions are essentially thermionic emission from the metal electrode, and in the absence of other effects, might be expected to have a Boltzmann $\exp(-\Phi_\mathrm{B}/k_\mathrm{B}T)$ dependence. What Schottky first explained is that there exists an attractive force between the electron injected from the cathode into the ferroelectric, and the other electrons remaining in the cathode, and that this attractive 'image' force has the net effect of lowering the barrier height Φ_B by an amount

$$\Delta\Phi = \left(\frac{eE}{4\pi\varepsilon}\right)^{1/2} , \tag{4.4}$$

as shown in Fig. 4.2. However, it is not readily apparent what value of field E is. It is certainly not the average value V/d, where V is the voltage drop across the whole ferroelectric film of thickness d. Sze states [131] that it is the maximum field in the junction region, which might be an order of

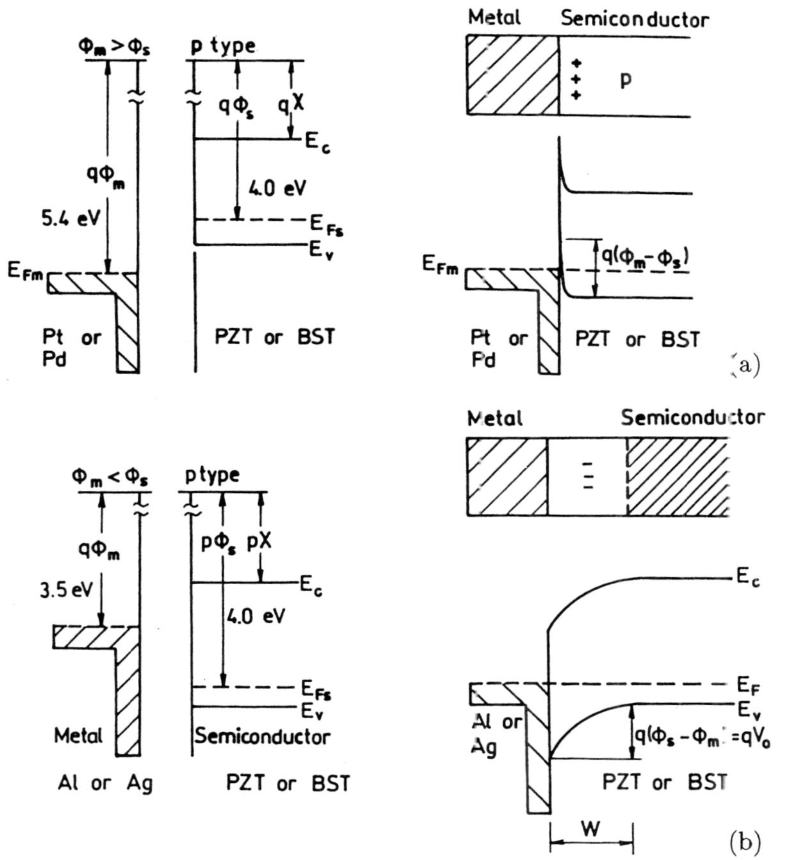

Fig. 4.1. Energy diagram for a metal electrode on a ferroelectric semiconductor, with an abrupt interface, at zero applied voltage. Energy (*vertical*) is plotted versus depth z into the film (*horizontal*). (**a**) p-type ferroelectric semiconductor on a metal (e.g., Pd, Pt) whose work function is larger than that of the ferroelectric. (**b**) p-type ferroelectric on a metal (e.g., Ag, Al) whose work function is smaller than that of the ferroelectric

magnitude larger than the average field $E = V/d$. Unfortunately, many publications on ferroelectric thin films assume $E = V/d$ and therefore calculate incorrectly (by an order of magnitude) $\Delta\Phi$. Similarly, because the electrons moving through the barrier spend a time in it short compared with the period of a long wavelength longitudinal optical phonon (wave number around 100 cm^{-1}), Frohlich interaction between injected electrons and phonons is very limited, and ε in (4.4) is ε_∞, the optical dielectric constant (equal to the square of the refractive index, around 5 6) and not the dc dielectric constant $\varepsilon_0 \approx 300$. We emphasize here that for PZT, BST, or SBT films, $\Delta\Phi$ is about 0.8 eV and slightly larger [135] than the 0.5 eV commonly encountered in Si

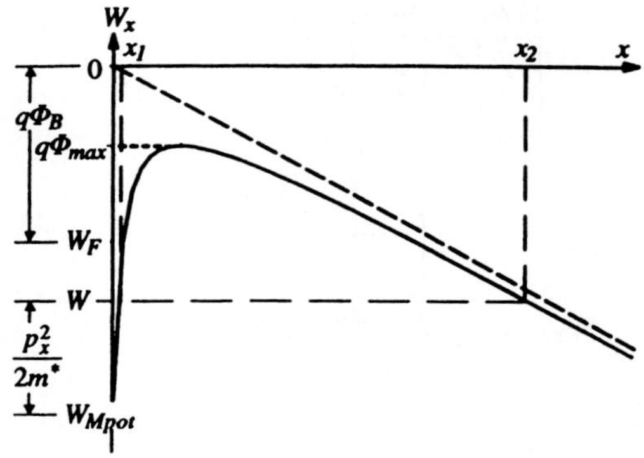

Fig. 4.2. Energy diagram for the band match-up in Fig. 4.1 but at an applied voltage of a few volts over 200–300 nm. This shows the reduction in Schottky barrier height $\Delta\Phi$ as well as the depletion width d_{d}. No surface traps (donors or acceptors) are included in this model

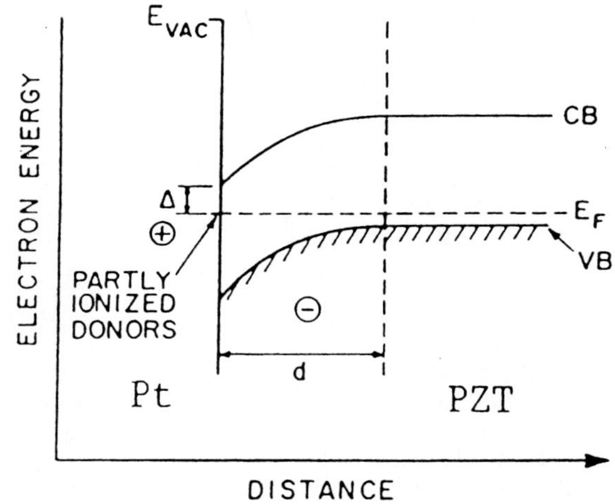

Fig. 4.3. Energy band diagram for the band match-up in Fig. 4.1 but with an n-type inversion near the interface due to surface traps (including oxygen vacancies). In this case, compared with Fig. 4.2, the depletion width can be quite different than in the defect-free case

devices. This correction is very large, since $\Delta\Phi$ is about 50% of Φ_{B} itself at 3 V.

BST, PZT, and SBT are all typically p-type semiconductors in bulk, if undoped [135]. This is because [136] there is a greater abundance of impurities such as Al, Na, and Mg in the starting growth materials than there are high-valence impurities, and because impurities are almost always substitutional in ABO_3 perovskites (not interstitial), with Na going in for Ba or Pb or Sr, Al^{+3} for Ti^{+4}, and Mg^{+2} for Ti^{+4}. Hence, *these are all acceptors, and as a result the ferroelectrics, at least in bulk, are generally all p-type semiconductors.* (This can be changed in thin films by surface donor states, which are apt to create n-type inversion layers but not convert the entire film to n-type.) Therefore at large distances z from the electrode interface the valence band is nearer the Fermi level than the conduction band is. However, near the electrode interface band bending produces a slight inversion (Fig. 4.3) so that the interface region is n-type. It is known [135] that diffusion of electrode materials into the semiconducting dielectric films can produce an n-type surface layer on a nominally p-type dielectric semiconductor, and this is the process that may occur in many ferroelectric memory films, including PZT/Pt, BST/Pt, and SBT/Pt. In all common ferroelectric thin film memory devices the conduction is predominantly due to electrons injected from the electrode (cathode), not due to holes (despite the nominal p-type character of the film). This point has been agreed to by all authors except Wouters et al. [137] and Tagantsev and Stolichnov [137]. The conclusion of Wouters et al. and subsequently of Tagantsev and Stolichnov apparently resulted from a misreading of the work of Dixit et al. [137], which actually is completely correct in this regard, and by the erroneous assumption that the Fermi level in PZT/Pt was pinned at mid-gap. The best way to verify an n-type inversion layer at the surface of a ferroelectric PZT or BST film (or at its electrode interface) is to measure Ti^{3+} concentration. Ti^{3+} is stable only in n-type materials. Care must be exercised, however, in that Ti^{3+} measurements in BST [138] via Auger techniques can show artifacts, according to Bardi et al. [138], produced by the Auger bombardment. The use of Ti^{3+} in this regard was first discussed by Merz and Anderson in 1955 and used to support a model of charge injection leading to interfacial space charge and fatigue [138]. In BST Scott and Ross reported [138] Ti^{3+} concentrations from the Pt interface down to about 110 nm (a third of the way through their BST film) about 10% of the Ti^{4+} concentration.

It is possible to prepare n-type ABO_3 perovskite ferroelectrics also. This occurs if the specimens are grown in reducing atmospheres. Oxygen loss at elevated temperatures is quenched into the room temperature structure. When an oxygen is lost, it exits the crystal as neutral O_2, leaving behind two electrons. The vacancy thus behaves as a donor, with a donor binding energy of around 1.0 eV. At room temperature many of these electrons at oxygen vacancy traps are thermally excited to form an n-type semiconductor. n-type perovskites usually look different from p-type; they are visibly darker.

Although acceptors and donors play important roles in the electrical and optical properties of ferroelectric titanates, tantalates, and niobates, there is little evidence for any important contribution from donor–acceptor (D–A) pairs. In this sense these materials are rather different from II–VI semiconductors.

4.2 Modifications to Schottky Theory for Ferroelectric Films

Both Schottky emission and Fowler–Nordheim tunneling are surface-limited currents. They depend upon the details of the potential barrier shape, height, and width at the ferroelectric–electrode interface. But they do not depend upon bulk characteristics of the film per se. Simmons has combined these two mechanisms and given detailed results, discussed further in Chap. 5, for the current density $J(V)$ as a function of applied voltage V. In general, Schottky emission will be dominant below 5 V (across 200–300 nm) and Fowler–Nordheim above. Even in the simplest case, as shown by Mark and Hartman [139], a film with Schottky interfaces and Poole–Frenkel conduction in its interior will have a slope in its current–voltage relationship (log J versus $V^{1/2}$) that is intermediate between the 2:1 ratio expected for pure Schottky and pure Poole–Frenkel conduction.

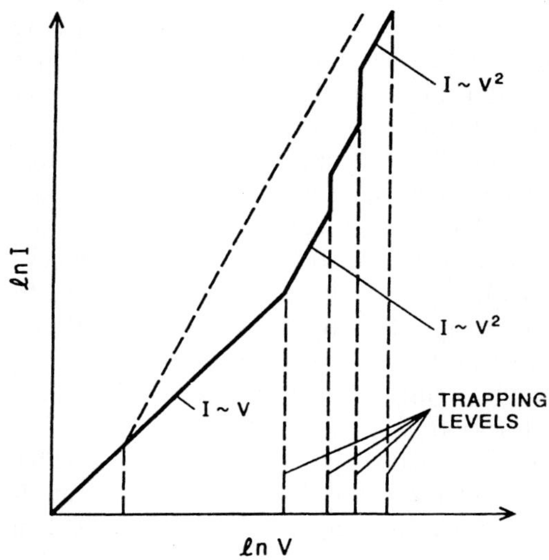

Fig. 4.4. Theoretical value of space-charge-limited current $J(V)$ versus voltage. Discontinuities indicate trap energies [151]

Unfortunately, we must also consider bulk scattering limitations upon $J(V)$, especially Poole–Frenkel scattering. Poole–Frenkel scattering has nearly the same temperature and voltage dependence for current density as does Schottky current, and the two are usually simultaneously present in the same samples. Li and Lu [139] were able to show in $BaTiO_3$ that there is a threshold in trap density above which Poole–Frenkel processes are the current-limiting ones. Combining Poole–Frenkel terms with Schottky and/or Fowler–Nordheim requires solution of an integral equation and generally has not been done for real ferroelectrics, but results to date are discussed further in Chap. 5. Qualitatively, the effect of Poole–Frenkel is easily separated from Schottky (and Fowler–Nordheim) by making top and bottom electrodes of different metals and measuring the asymmetry in leakage currents with positive and negative polarity. Poole–Frenkel is symmetric with respect to polarity under these conditions. The actual data for PZT or BST (Fig. 4.4) are very asymmetric.

4.3 Charge Injection

In a rather different context, namely the development of nerve-stimulating contacts in medical physics, Robblee and her coworkers have shown in detail that Ir/IrO_2 are the electrodes of choice for any oxygen-containing ambients, and that platinum is non-optimum [140, 141, 142, 143, 144, 145, 146, 147]. The basic reason is that even at room temperature, repetitive voltage cycling of Pt electrodes produces platinum oxides (although some of my colleagues claim this is not true [148]). This leads to changes in the Pt/ferroelectric barrier height, to increased leakage currents, and to time-dependent breakdown. In contrast with Pt, Ir oxidizes quite reversibly, by changing the Ir valence. Thus with no oxygen transport, Ir can oxidize and reduce as voltage is cycled. This acts as a buffer to prevent excess charge injection into the dielectric and thus prevents breakdown. This is also the reason Ir/IrO_2 prevents fatigue in PZT films, that is, by buffering the charge injected from the metal electrodes when bipolar switching is undergone. Robblee's work was quite quantitative. She pointed out that up to $\rho = 20\ \mu C/cm^2$ of charge density could be injected through a double layer of Ir/IrO_2, but that injection densities greater than that ($2P_r > 20$) would lead to dielectric breakdown or Faradic reactions (oxidation or reduction of some ion in the ferroelectric due to charge injection) – the latter causing fatigue. This is unimportant in most ferroelectrics, such as SBT, because $P_r < 10\ \mu C/cm^2$. But it is very important in PZT. It is the reason that 'good' PZT with $P_r > 20\ \mu C/cm^2$ fatigues rapidly but 'bad' PZT with P_r around $4\ \mu C/cm^2$ does not fatigue at all. And similar arguments apply to SBTN, as shown by Noguchi et al. [149] and by the present author [149], as shown in Fig. 4.5.

4.4 Space-Charge-Limited Currents (SCLC)

We have outlined briefly above the major mechanisms of leakage current in ferroelectric films – Schottky, Poole–Frenkel, Fowler–Nordheim. However, when one of these, or any combination of them, reaches a certain threshold level, such that the current is no longer limited by details of the electrode interface, the current properties take on very special characteristics – they

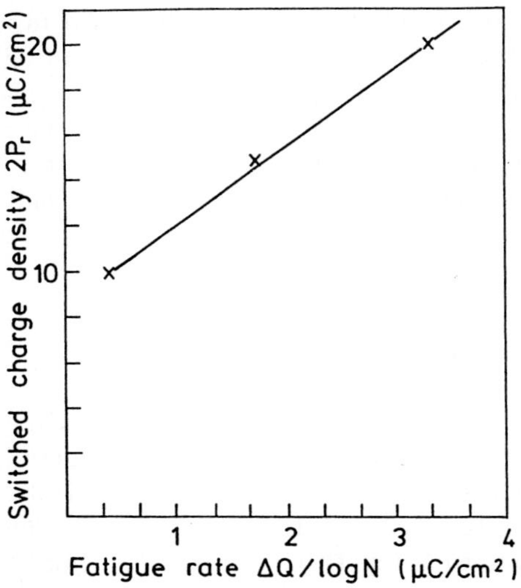

Fig. 4.5. Switched charge density in SBTN versus fatigue rate

become limited by the injected space charge in the ferroelectric. The first direct observation of injected space charge and of space-charge-limited currents was in II–VIs by Smith and Rose on CdS and soon after, by Ruppel on ZnS [150]. It is useful to keep in mind that space-charge-limited current densities in solids are always orders of magnitude less than those in vacuum tubes (where a typical value is 1.0 kA/cm^2).

The physics of this phenomenon are nearly identical to the vacuum tube case, first analyzed in the early part of this century. At very low voltages, the current in a vacuum tube is limited by thermionic emission from the cathode. The space between the cathode and anode plays no important role. But as the discharge current is increased, the gap between cathode and anode becomes filled with electrons. This injected charge retards the transmission of subsequent electrodes; it is called *space charge* because it literally fills the space between cathode and anode. At very high currents, it is the current-limiting effect, and the exact work function of the cathode, and other thermionic emission details, become negligible. This regime is called *space-charge-limited current.* In this regime the current varies not ohmically as V, but as V^2; and at constant voltage V, it varies not as $1/d$ (ohmic), but as $1/d^3$. This

$$J = \text{constant}\ \frac{V^2}{d^3} \tag{4.5}$$

dependence is known as the Mott–Gurney law (similar to Child's law for vacuum tubes) and also applies to ferroelectrics, provided one inserts the dielectric constant ε in the right places and makes certain other minor corrections. Equation (4.5) is derived under the assumption that the drift mobility is field independent. This should be a good assumption in ferroelectric oxides up to a few hundred kV/cm. We note parenthetically that for a point contact, (4.5) has

$$J = \text{constant } V^{3/2} \tag{4.6}$$

rather than V^2. Such a leakage current dependence does not seem to have been reported in the literature for ferroelectric thin films, although it has been reported in II–VIs [151].

Before discussing the voltage exponents that appear in (4.5,4.6) and in (4.11) and (4.13–4.16) below, let me emphasize that experimentally determining this value is in itself insufficient to determine the physical mechanism of conduction. Generally the exponent characterizing the thickness dependence is also required. The reason is that more than one mechanism can give rise to the same voltage dependence. For example, vertical jumps in a $J(V)$ curve can arise from electrical breakdown or from field-induced emptying of filled traps, or from impact ionization of filled traps (impact ionization across the bandgap is negligible in these ferroelectrics because the bandgap is too large). As another example, a V^2-dependence of $J(V)$ can arise from double injection or from single injection in the trap-filled limit. Therefore it is generally not prudent to speculate about detailed conduction mechanisms unless a number of self-consistency checks are afforded by the data. Generally this will require $J(d)$, $J(T)$, and $J(V)$ data.

For ferroelectric PZT and BiT the predicted dependence of $J(V)$ in (4.5,4.6) is shown in Fig. 4.4 [151] and confirmed experimentally [152] in Fig. 4.6a,b. $J(d)$ in (4.5) is confirmed experimentally for PZT [153, 154] in Fig. 4.6c. In actual practice, the simple form of (4.5) must be replaced by

$$J(E) = A_0 E + B_0 (E - E_0)^2 , \tag{4.7}$$

where $eE_0 d$ is a trap energy (shown in Fig. 4.4), A_0 is an ohmic, bulk term, and B_0 is a surface-related film term.

Similarly, the dependence upon film thickness d in (4.5) is more accurately given not by d, but by $(d-2a')$, where a' is an *accommodation length* or *mutual diffusion length* near the electrodes:

$$J(d) = \text{constant } (d - 2a')^{-3} . \tag{4.8}$$

In Si devices, Mayer finds [155] a'/d is about 10%, and we estimate the same ratio in ferroelectric PZT films ($a' = 20$ nm, $d = 200$ nm).

If we express (4.5) more precisely for ferroelectric films, we find that there are a number of corrections. These corrections will in general be temperature dependent. The theory is best parametrized by a constant l which is defined as

$$l = \frac{T_\text{f}}{T} , \tag{4.9}$$

where T_f is related to the density of trap states per unit energy near the Fermi surface of the ferroelectric [156, 157] and T_f is found to be [158] about 390 K

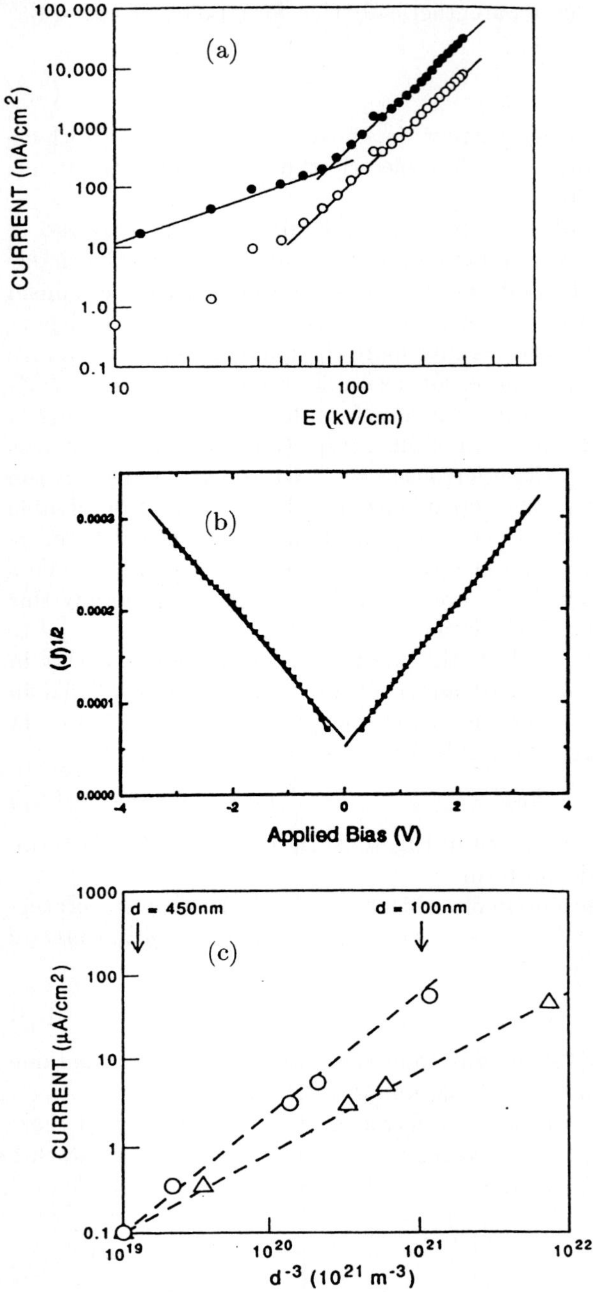

Fig. 4.6. (**a**) Experimental confirmation of (4.5) for $J(V)$ [152] in PZT, (**b**) for BiT. (**c**) Experimental confirmation of (4.5) for $J(d)$ [153]

for SBT films. Thus, l is nearly but not exactly unity for room-temperature measurements. The temperature dependence of (4.5) can be approximated by

$$J(V, T) = \text{constant}\ \frac{V}{d}\left(\frac{V}{d^2}\right)^{T^*/T} \tag{4.10}$$

which satisfies the BST data shown in Fig. 4.7. T^* is 310 ± 20 K in all these fits, so that the difference between (4.10) and the Mott–Gurney law (4.5) is small (i.e., T^*/T is near unity at room temperature operation). However, it is important that (4.10) permits $J(V)$ to be calculated at any T if it is known at one temperature.

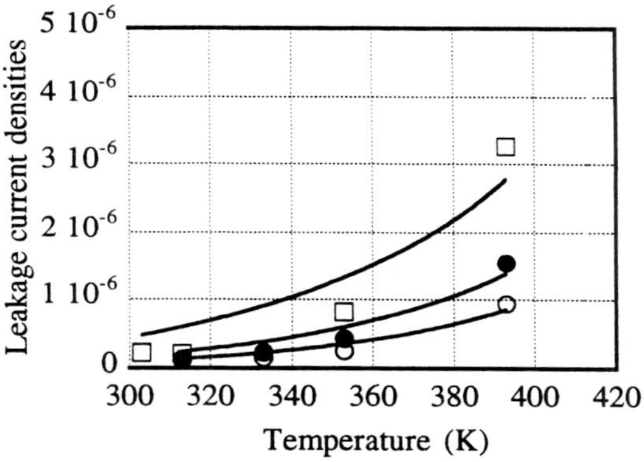

Fig. 4.7. Experimental confirmation of (4.10) in BST

The result of this consideration is that (4.5) should be replaced [159] with

$$J(V, d) = A_0 \frac{V^{(l+1)(1-ma'/d)}}{(d-2a')^{2l+1}}\,, \tag{4.11}$$

where m is an empirical constant between 2 and 3.

Once one is known to be dealing with the SCLC regime, a number of useful relations can be employed. Firstly, it is necessary to decide whether one is in the regime of charge neutrality. Waser has assumed this to be true of most PZT data [160], but Brennan [161, 162] has assumed to the contrary that there is a non-equilibrium distribution of net charge. In this regime one must substitute Poisson's equation

$$n - p = \frac{e}{\varepsilon}\frac{\mathrm{d}E}{\mathrm{d}z} \tag{4.12}$$

for the equation of charge neutrality $n = p$. This is in fact the SCLC regime, which is encountered with increasing voltage as a kind of insulator–metal transition. Lampert (p.284) discusses carefully the region of validity of this neutrality assumption. In the low-voltage insulating (or semi-conducting) regime, one finds [157] [ignoring for simplicity the corrections to (4.5) given in (4.7) and (4.8)]

$$J(V,\, d) = \frac{9}{8}\, e\, t_{\mathrm{R}}\, \mu_{\mathrm{n}}\, \mu_{\mathrm{p}}\, N\, V^2 d^{-3}\ , \tag{4.13}$$

where t_{R} is a recombination time, μ_{n} and μ_{p} are the mobility of electrons and holes, and N is the density of recombination centres. This may be contrasted with the equation for the high-voltage regime:

$$J(V, d) = \frac{125}{18}\, \varepsilon\, t_{\mathrm{R}}\, \mu_{\mathrm{n}}\, \mu_{\mathrm{p}}\, V^3 d^{-5}\ , \tag{4.14}$$

where we note that the dielectric constant ε enters (4.14) but not (4.13).

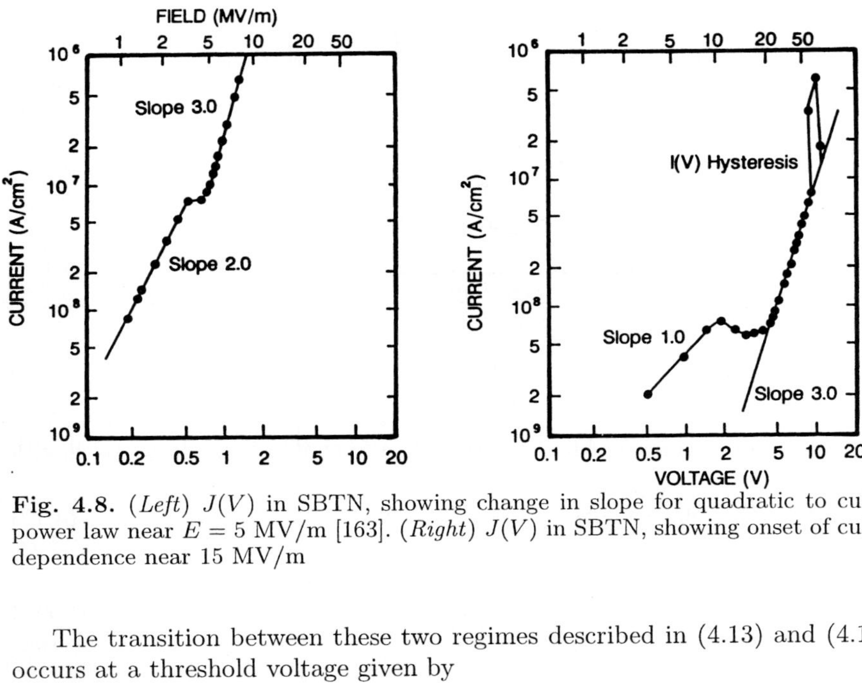

Fig. 4.8. (*Left*) $J(V)$ in SBTN, showing change in slope for quadratic to cubic power law near $E = 5$ MV/m [163]. (*Right*) $J(V)$ in SBTN, showing onset of cubic dependence near 15 MV/m

The transition between these two regimes described in (4.13) and (4.14) occurs at a threshold voltage given by

$$V_{\mathrm{th}} = \mathrm{constant}\ \frac{d^2}{2\mu_{\mathrm{n}}\mu_{\mathrm{p}}}\ , \tag{4.15}$$

which according to Figs. 4.8a and b, may occur [163] at anywhere from 2 to 15 MV/m in SBT and 10 MV/m in PZT [164].

We note parenthetically that for a completely uniform distribution of traps (uniform with regard to energy), the power law dependences of $J(V)$ discussed above go over to an exponential dependence

$$J(V, d) = e\,\mu\, n_{\mathrm{e}}\,\frac{V}{L}\,\exp\frac{\alpha V}{T d^2}\,. \tag{4.16}$$

This situation does not seem to be a good approximation to real ferroelectric films.

Space-charge-limited currents have other characteristics of importance, including the fact that they are not monotonic in time but rise to a maximum and then decay, after the voltage pulse is delivered. This was shown clearly by Many [165]. The problem experimentally is that this maximum is reached at very short times and has generally been unmeasurable. However, in one particular and unusual case, transient space-charge-limited currents from ions (not electrons) have been demonstrated very clearly in ABO_3 ferroelectrics, which is the very recent work from Motorola [166].

4.5 Negative Resistivity

We note in Fig. 4.9 that there is a voltage region where $\mathrm{d}J/\mathrm{d}V < 0$. This region of negative resistivity can arise due to either double-injection (electrons from cathode and holes from anode) or trap filling and trap emptying. The latter is well known in bulk PZT [167]; see also Kroger et al. for earlier studies [167]. The former – double injection – is known not to occur in $BaTiO_3$ except under optical illumination [168].

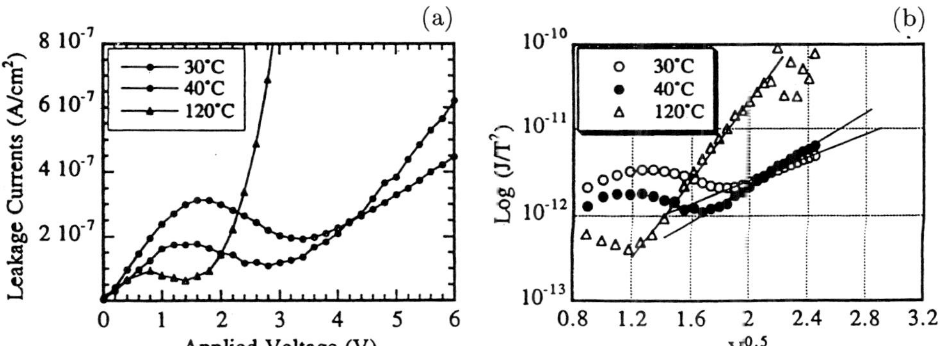

Fig. 4.9. (**a**) Negative resistivity in Bi-excess SBT on Pt electrodes, (**b**) in SBT on Bi electrodes

Independently of these details, one can estimate the total trap density within $k_{\mathrm{B}}T$ of the ferroelectric Fermi surface by noting the voltage ΔV required to double the current. In PZT films of around $d = 250$ nm this is 10 V, and hence [169]

$$N_{\mathrm{t}} = \frac{\Delta V\, C}{\varepsilon} = 1.1 \times 10^{19}\ \mathrm{cm}^{-3}\,, \tag{4.17}$$

assuming $\varepsilon = 1300$. We shall see that trap densities, both theoretical and experimental, vary by several orders of magnitude, from 10^{18} to 10^{21} cm^{-3}. The reason is that bulk trap densities in ferroelectric oxides are moderate (of order 10^{18}) whereas surface trap densities are very large (10^{20}–10^{21}), so estimates made in different ways may weight more heavily surface or bulk properties.

The number in (4.17) is significantly smaller than the total trap density of 10^{21} cm^{-3} from thermally stimulated current flow [170] or from earlier electrochemical studies [171], but it agrees with the value from two other estimates. Firstly, the ratio of free electrons to trapped electrons in an SCLC device can be determined from the voltage at which the linear (ohmic) and quadratic (SCLC) currents intersect. In PZT this is about 1.5 V. (One must be very careful in using the apparent Ohm's law region of $J(V)$ data for these purposes; such an apparently ohmic region often arises from surface currents if guard rings are not used.) The relationship between this threshold voltage and the number of trapped electrons N_{tr} is

$$N_{\mathrm{tr}} = 5.5 \times 10^5 V_{\mathrm{th}} \varepsilon d^{-2} = 2.7 \times 10^{18}\ \mathrm{cm}^{-3}\ , \tag{4.18}$$

which must be less than N_{t} (since not all the electrons are trapped); that is, (4.17) and (4.18) indicate that 25% of the electrons in one PZT sample were trapped. A second estimate is given by the voltage V' at which $J(V)$ displays a nearly vertical jump (e.g., 12 V in SBTN 180 nm thick). V' is related to the density of trapped holes p_{tr} by

$$p_{\mathrm{tr}} = 1.1 \times 10^6 \varepsilon V' d^{-2} = 3.3 \times 10^{18}\ \mathrm{cm}^{-3}\ . \tag{4.19}$$

Thus for PZT all estimates based upon SCLC calculations yield an estimate of a few times 10^{18} cm^{-3} for the trapped electron or hole concentration in bulk and a few times 10^{21} cm^{-3} trap concentration near the surface. The bulk number agrees with experiments of Mihara et al. [172], based on grain-boundary depletion models [173], and with independent results of Dey and Zuleeg [174]. Note that as initially shown by Klimov [175], only about 5% of the oxygen vacancies trap electrons; the others are compensated by lead vacancies in PZT.

We end this chapter on a somewhat discouraging note. Jo et al. [176] have shown that nominally identical ferroelectric thin films of bismuth titanate (BiT) can manifest Poole–Frenkel, ohmic, space-charge-limited, or Fowler–Nordheim leakage currents, depending upon the electrode and substrate used and the annealing process used. They find that BiT on glass with an oxygen anneal is space-charge-limited. BiT on glass annealed in nitrogen is Poole–Frenkel. Yet BiT on LSCO (Lanthanum Strontium Cobaltous Oxide) electrodes are ohmic if the LSCO is epitaxial, but space-charge-limited if the LSCO is grain-oriented; and it is Poole–Frenkel if the LSCO is polycrystalline! BiT/LSCO goes over to Fowler–Nordheim at about 90 kV/cm, rather lower than in SBT (around 400 kV/cm).

Not everyone believes that leakage currents in ferroelectric thin films can be described by Schottky barrier heights, electrode work functions, ferroelectric electron affinities, and the other subtle parameters in band structure match-ups at the interface. Toshiba has produced two recent papers [177, 178] in which they argue that these factors are negligible or irrelevant and that instead, for barium titanate heteroepitaxy on platinized MgO, on $SrRuO_3$, or on $SrTiO_3$:Nb electrodes, the leakage current is dominated by lattice misfit and strain. I know of no direct mechanism whereby lattice strain can cause an increase of orders of magnitude in the leakage current of a wide bandgap semiconductor such as $BaTiO_3$, and indeed these authors suggest none. Indirectly, however, lattice strain can shift the valence and conduction bands in the ferroelectric, relative to the metal electrode work function, to make the junction change from Schottky-blocking to ohmic. However, in the Toshiba work their own data show that they are not even measuring leakage current. At their 5 s/V measuring rate they apparently measured only polarization relaxation. This was made explicitly obvious by the fact that the current measured $J(V)$ was completely different for increasing V and for decreasing V, thus showing that their system was never in steady state. Their data appear completely 'ohmic' (J linear in V) above a large voltage threshold ($V \approx 20$ V) and they argue that there is an ohmic conduction region above a low-voltage Schottky regime. It is far more likely that they are seeing the onset of Fowler–Nordheim tunneling, which is known to set in in ABO_3 perovskite films at high voltages. Their data are an important example for students and those new to the field. Never use $J(V)$ data that are different for increasing and decreasing voltages! These are clearly not in the steady state regime of true leakage current; they indicate either TDDB at long dwell times or relaxational currents at very short dwell times. Unfortunately, as shown by Waser, the steady state leakage current regime for pure $SrTiO_3$ or $BaTiO_3$ is > 300 s at ambient temperatures.

One curious aspect of the Toshiba work is not impossible. They find an ohmic, linear $J(V)$ region at voltages above a faster V-dependence. Such situations can easily occur if there are two sets of traps, of which one lies above the Fermi energy in the ferroelectric and one below, as shown in Lampert's text (p.23).

Parenthetically, in any discussion of negative resistivity of ferroelectric thin films, it must be emphasized that this phenomenon is not related to negative resistivities in Gunn oscillators (GaAs devices). Negative differential resistivity in thin insulating films has been studied carefully for forty years and is not controversial [179, 180, 181, 182, 183, 184]. As Klein stated [185]: "It has been observed that when thin insulating films between 20 and 300 nm thick have undergone a forming process, the conduction process manifests voltage-controlled negative resistance." Particularly quantitative studies of negative differential resistivity have been published for Au^{-2} traps in n-Ge by Pratt and Ridley [185].

Some authors have noted that BST films occasionally exhibit highly non-linear $J(V)$ behaviour at high voltages that are like *varistors*. It is my contention that this is primarily the onset of Fowler–Nordheim conduction. In the well-studied ZnO varistors [186], it is the stability of the intergranular phase that produces varistor effects; and these are highly sensitive to heat treatment in oxygen. Oxygen transport within BST ceramics is certainly important, but it is not yet established whether the $J(V)$ mechanisms in ZnO and BST are the same in this varistor regime.

It is well known that negative resistivity is unstable against filamentation [187]. Therefore negative resistivity can be used by ferroelectric memory test engineers as a warning sign, a precursor to filamentary electrical breakdown. In Si devices this filamentation does not occur until a current threshold of around 10 mA. Single discrete filaments could be observed in p-Si:In at such values [187]. But negative resistivity in PZT [167] or in SBTN [163] occurs at lower current densities, in the $\mu A/cm^2$ range. In a fine-grained ceramic, filamentation will involve quasi-reversible breakdown at grain boundaries and a distinct hysteresis in the J versus V curves at high voltages [163]. These effects were first reported in II–VI semiconductors by Nicoll and by Bube [188].

Negative resistivity does not usually occur in SBT, but it is easily induced by adding Bi excess. These form surface traps near 2.2 eV, and the emptying of these traps with applied voltage causes the negative resistivity shown in Fig. 4.9a at different temperatures. SBT on Bi electrodes exhibits the same effect (Fig. 4.9b).

5. Capacitance–Voltage Data: C(V)

In a ferroelectric thin film the capacitance depends strongly upon voltage. Expressed in terms of electric field E this voltage dependence arises from two unrelated sources. The first is the nonlinear dependence of dielectric constant upon field $\varepsilon(E)$. The second is the field dependence of the width $d(E)$ over which most of the voltage drop occurs. The general form is

$$C(E) = \frac{\varepsilon(E)}{d(E)} \; . \tag{5.1}$$

Typically scientists and engineers working on ferroelectric thin films for memory devices Dynamic Random Access Memory capacitors (DRAMs), or Non-Volatile Random Access Memories (NV-FRAMs) in which the ferroelectric actually stores the information, have assumed that all the field dependence is in $d(E)$ and have identified $d(E)$ with the depletion width in the ferroelectric (most ferroelectrics used in memory applications are wide bandgap p-type semiconductors) [189, 190, 191, 192, 193, 194 195, 196]. For those memory engineers whose background was with silicon devices, this was a natural assumption, since voltage dependent depletion widths are dominant in analogous silicon device phenomena. Credit should be given to Evans [196] for first correctly determining that PZT had a space charge layer of 20 nm at the electrode interface, in contrast to the Sandia claim that it did not exist [197]. Not only ceramics but single crystals of semiconducting materials (including ferroelectrics) exhibit inhomogeneous space charge regions surrounding imperfections, as shown by Weisberg [198]. These can be studied and in fact removed by optical photoexcitation, as detailed by Nelson et al. [199]. Inhomogeneous space charge often results in temperature dependence of Hall mobilities. Thermal quenching typically results in a mobility drop of 10^4 or more over a narrow temperature range, as described by Bube [200].

At the other limiting approximation, some scientists have found evidence in their titanate ferroelectrics that [201 202] the primary effect in $C(V)$ data was due to the nonlinear permittivity $\varepsilon(E)$, with the implication that d is constant, independent of field, and is equal to the true film thickness [203]. The latter assumption means that the semiconducting ferroelectric films are fully depleted. Full depletion, in turn, means that the conduction and valence bands tilt rather than bend (Fig. 5.1) at the electrode–ferroelectric interface, and via Poisson's equation

$$\frac{\mathrm{d}\Phi}{\mathrm{d}x} = \rho(x) \tag{5.2}$$

in one dimension, that there is no space charge. In fact there is a simple and analytic relationship between $C(V)$ and $\rho(x)$, the depth dependence of the space charge near the interface [204, 205]: if

$$C(V) = \text{constant } V^{-1/n} , \tag{5.3}$$

and

$$\rho(x) = \text{constant } x^{-m} , \tag{5.4}$$

then $n = m+2$. Thus, a quadratic space charge dependence upon depth gives an inverse quartic voltage dependence of capacitance (as in an abrupt pn-junction), and a constant space charge dependence yields an inverse quadratic voltage (or field) dependence for capacitance C.

We note parenthetically that the metal electrode–ferroelectric interface behaves like a pn-junction in some respects, including its leakage current and $C(V)$ behaviour. The grain boundaries and electrode interface behave like back-to-back pn-junctions due to depletion space charge layers and surface donor states; these will produce $J(V)$ current behaviour like reverse-biased pn-junctions. However, the reverse-bias algebra for a metal–semiconductor interface is considerably more complicated than that of a pn-junction, because the current depends strongly on the width of the space-charge region. After fatigue, the oxygen vacancies accumulate at the cathode. This produces enhanced p-type conductivity in the anode region, and hence the film behaves like a forward-biased pn-junction leading to degradation and resistance breakdown (TDDB). In either case, reverse- or forward-bias, the junction depletion width is relative to the carrier mean free path, and one should keep in mind that the mean free path in a ferroelectric oxide at room temperature is only about 0.1 nm. Therefore even good PZT/Pt junctions will be 'wide' compared to the mean free path.

In practice n is often taken as an adjustable empirical parameter in ferroelectric thin films [206] of order 3.0, without claiming a precise physical model.

At present these different approaches have produced confusing inconsistencies regarding the depletion widths d_{d} in the important class of DRAM capacitors consisting of thin films of $SrTiO_3$ or its BST ($Ba_xSr_{1-x}TiO_3$) perovskite isomorphs. Six different publications from respected groups give widely different values for the apparent depletion widths in their BST specimens. [203, 207] conclude that $d_{\mathrm{d}} \gg 50$ or $\gg 200$ nm (the film thicknesses d) and hence that the films are fully depleted. But [193, 208, 209, 210, 211] give rather different estimates of d_{d} from 1 nm to < 13 nm, with the most quantitative estimate [211] 6.4 nm at $V = 0$ applied voltage (increasing to 38 nm at 8 V but still much less than the film thickness).

It is a primary aim of the present chapter to reconcile these seemingly irreconcilable differences. In reviewing the literature we do not include all

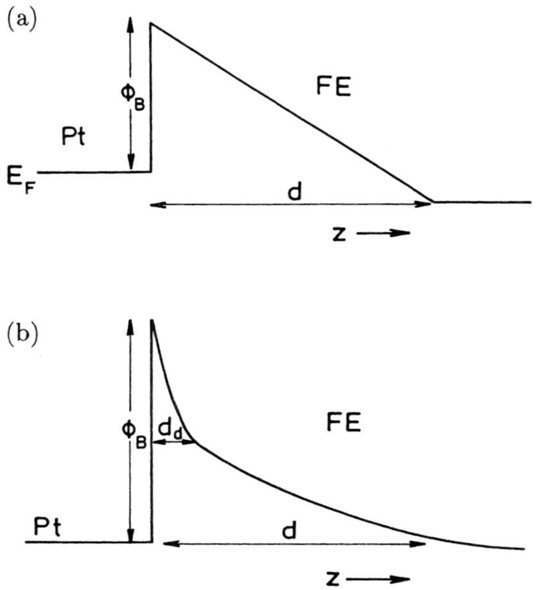

Fig. 5.1. Potential versus film depth z near the metal electrode–ferroelectric interface in a thin-film memory device. (**a**) Assuming band tilting and no space charge in a fully depleted film. (**b**) Assuming that there is band-bending near the interface due, for example, to oxygen depletion. Fowler–Nordheim tunneling is likely in (**b**) but not (**a**)

estimates of d_{d}. Some are thought to be unreliable. But neither do we fall into the lame excuse that work that disagrees with our own is wrong or arises from inferior materials. Rather, we will try below to reconcile the apparent disagreements for the most part via a model in which the conduction band structure in the ferroelectric semiconductor near the metal electrode interface is rather complicated (due in part to the fact that the defect density is [195] a few times 10^{20} cm^{-3} near the interface, compared [191, 206] with a few times 10^{18} cm^{-3} in the film interior). This produces a band structure in applied fields which is neither simple band tilting (Fig. 5.1a) nor simple Schottky structure. Our model (Fig. 5.1b) resembles the silhouette of the Matterhorn, and we note coincidentally that this paper was conceptualized in an outing above Zermatt. In this model Fowler–Nordheim tunneling can occur through a thin (3–7 nm) peak in the conduction band near the electrode. But the low voltage leakage current is exactly ohmic, the high-field leakage current is trap-dominated (Frenkel–Poole) by the interior of the film, and the polarity dependence of the current is still Schottky-like and demonstrably different from Frenkel–Poole. Although this model seems to combine elements of Schottky, Frenkel–Poole, and Fowler–Nordheim conduction – that is, both bulk-limited and interface-limited conduction, it is neither new nor ad hoc. It derives directly from the work of Simmons [212, 213], Geppert [214], and Many [215], and earlier work by Adirovich [216] and Stoeckmann [217], and has been used by others as recently as 1997 [218, 219]. Because it is most easily tested via careful measurements of leakage current, those are emphasized in the present paper. However, it also has significant implications for UPS and XPS data and band matchup models from them for the electrode-ferroelectric interfaces, particularly regarding image field correction terms (for which one needs to know whether the voltage drop is uniform across the whole ferroelectric film). This is discussed in the latter part of the present paper.

5.1 Aspects in Favor of a Thin Depletion Layer

5.1.1 Richardson Coefficient A^{**}

When the standard Schottky equation (5.5) below is fitted to leakage current data in BST thin films, it is found that the dependences upon voltage V and temperature T are generally satisfied [220]. However, the overall magnitude, determined by the prefactor A^{**} (the modified Richardson coefficient) is much too low experimentally [198]. The discrepancy is as great as 10^6 at a few volts over 150–200 nm. This is, however, exactly the factor predicted by Geppert [214] for the case in which the Schottky dependence must be modified to incorporate Poole–Frenkel and other bulk scattering mechanisms for electrons in the interior of the film, as shown in Figs. 5.2–5.7 below. Thus this agreement of the effective Richardson coefficient A^{**} predicted from a thin depletion width and thin Schottky barrier at the surface and significant Poole–Frenkel sc attering within the film's interior with experimental data is a strong argument against models that have homogeneous band tilting and no surface depletion.

$$J = A^{**} T^2 \exp\left(-\frac{e\Phi_\mathrm{B}}{k_\mathrm{B}T}\right) \exp\left[\frac{e}{k_\mathrm{B}T}\left(\frac{eE}{4\pi\varepsilon}\right)^{1/2}\right] , \tag{5.5}$$

where Φ_B is the Schottky barrier height and A^{**} the modified Richardson coefficient. Most ferroelectric thin film capacitors exhibit the Schottky behaviour given in (5.5). That for BST in Fig. 5.2 is classic: the logarithm of leakage current $\log J(V)$ exhibits a perfect square root curve. Fitting $J(V, T)$ data to this formula usually gives very accurate values of the Schottky barrier height Φ_B, as shown for BST/Pt in Fig. 5.3 and for SBT/Pt in Fig. 5.4, even when it gives completely ridiculous values of A^{**}. This is discussed further on in the text. However, in PZT/Pt this procedure gives two different values of activation energies (or barrier heights) – about 1.5 eV at high temperatures and 0.2 eV at low temperatures (see Fig. 5.5. As Waser has pointed out, this is probably due to the failure of the system to reach steady state in the dwell

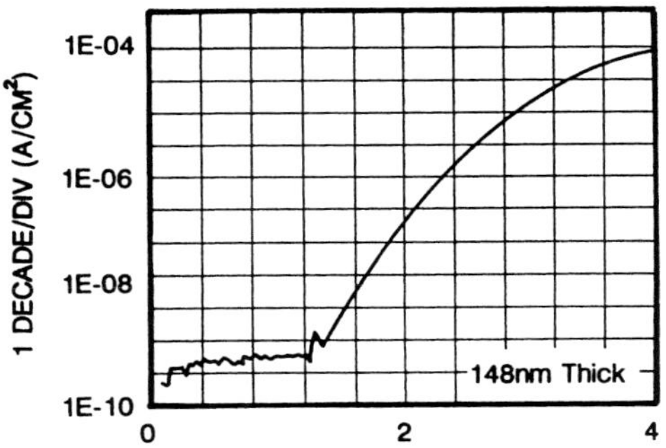

Fig. 5.2. Schottky behaviour of BST/Pt leakage currents. J versus voltage

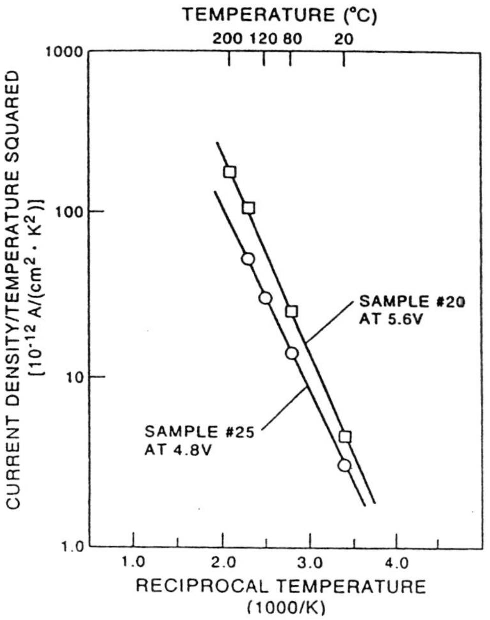

Fig. 5.3. Schottky barrier height in BST/Pt determined from leakage current $J(V, T)$

time (around 5 s) of the experiments at $T < 500$ K, including the room temperature measurements, so that the real barrier height in PZT/Pt is indeed approximately 1.5 eV at low applied voltages (i.e., neglecting image forces).

We note parenthetically that the bending over of the $J(V)$ curve in Fig. 5.6, due largely to the modifications to the Schottky formula suggested by Geppert and by Simmons, can also arise in large measure simply from the

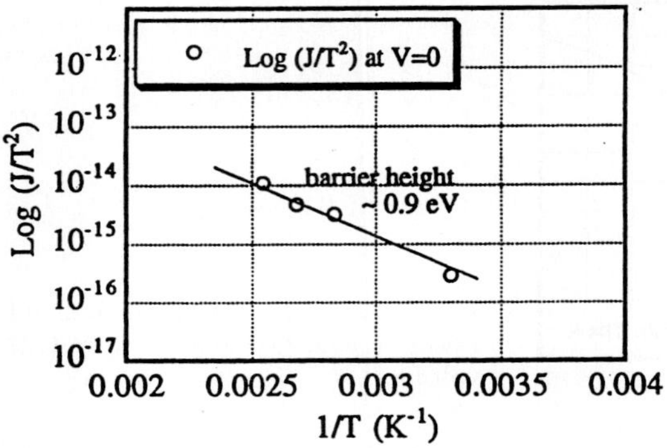

Fig. 5.4. Schottky barrier height in SBT/Pt determined from leakage current $J(V, T)$

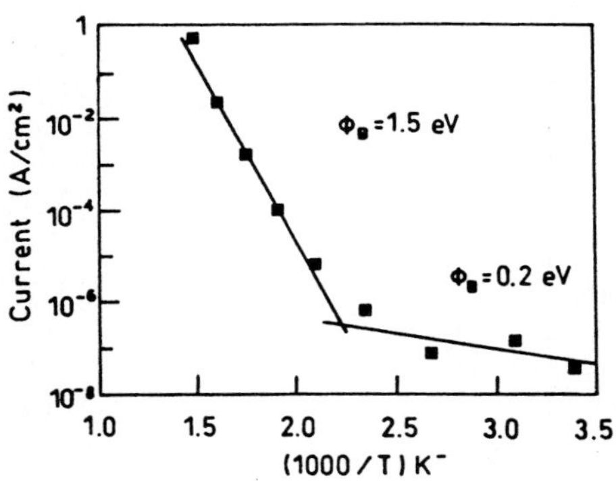

Fig. 5.5. Barrier heights inferred from leakage current data in PZT/Pt

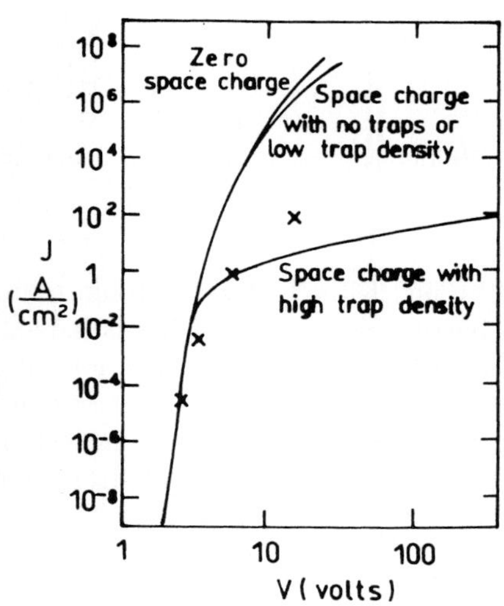

Fig. 5.6. Leakage current J versus applied voltage V from Geppert [214], showing Schottky-like behaviour but a reduction of around 10^6 due to Poole–Frenkel processes in the interior of the film

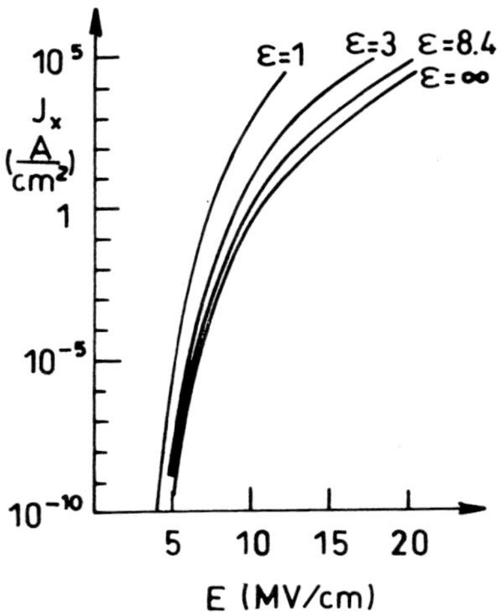

Fig. 5.7. Meyerhofer [214] calculates a reduction factor of around 10^6 due simply to image force corrections, even in the absence of Poole–Frenkel scattering

image force correction. As first shown by Meyerhofer [214], and reproduced in Fig. 5.7, this can produce as much as one order of magnitude reduction in J (i.e., in effective Richardson coefficient) for every one V/μm field, in agreement with BST experiments.

Equation (5.5) is derived assuming that the electron mean free path λ is greater than the ferroelectric film thickness d and much greater than the electrode–dielectric interface thickness. However, the mean free path λ is only about 0.1 nm in a ferroelectric oxide. Under these conditions Simmons showed [212] that (5.5) must be relaced by

$$J = \alpha T^{3/2} E \mu \left(\frac{m^*}{m}\right)^{3/2} \exp\left(-\frac{\Phi}{k_B T}\right) \exp\left(\beta E^{1/2}\right) , \qquad (5.6)$$

an equation that works very well for BST [212]. We note parenthetically that when used by Zafar et al. [212] to fit BST data, (5.6) was employed with the assumption that the field E is the average V/d in the film. However, we emphasize that this assumption, tantamount to full depletion (probably wrong) plays no important role in their results – it primarily changes the numerical value of the mobility fitted to the data. Note however that the coefficient β is not freely adjustable and is basically $\varepsilon^{-1/2}$. Therefore if E is very different from V/d, an error would be apparent in the fitted value of ε, which is not the case.

Combining (5.5) with Poole–Frenkel currents through the film's interior and Fowler–Nordheim tunneling

$$J = \frac{e^2}{8\pi h\Phi} E^2 \exp \frac{(32m^*)^{1/2}(e\Phi)^{3/2}}{3heE} \exp[-f(\Delta\Phi)] , \qquad (5.7)$$

through the 'tip' (Fig. 6.1b) of the Schottky barrier is a formidable undertaking, because it involves solution of a very nonlinear integral equation. (Here

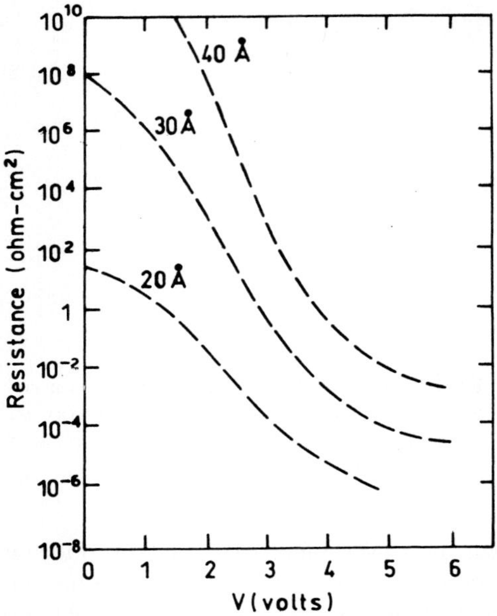

Fig. 5.8. Reduction in effective resistivity of a metal–ferroelectric interface with increasing voltage [212]

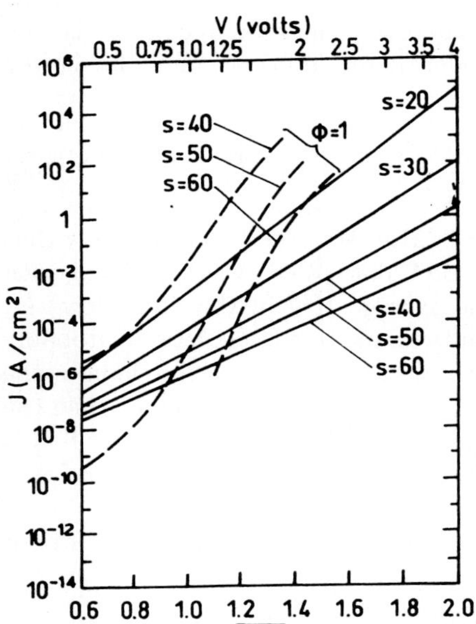

Fig. 5.9. Leakage current J versus voltage at a modified Schottky barrier [213], showing the threshold voltages at which Fowler–Nordheim tunneling exceeds thermionic Schottky emission. Note that this occurs at depletion widths of order 4–5 nm, as shown schematically in Fig. 5.1b

m^* is the electron effective mass and f an elliptic integral. It is useful to point out that m^* for electrons in $SrTiO_3$ or BST is $m^* = 6.00m_e$, the free electron mass, and also that the polaron coupling coefficient α for the lowest energy $q = 0$ phonon branch is 4.5 [214].) However, Simmons [212, 213] has already combined these numerically (Figs. 5.8, 5.9) and the addition of the Poole–Frenkel term can also be done numerically, as initiated by Geppert [214].

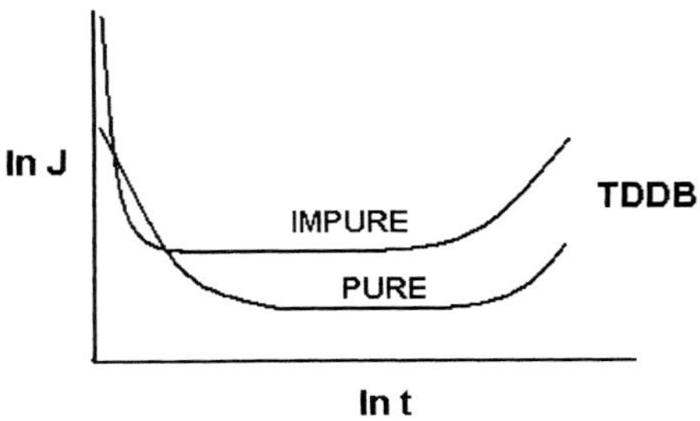

Fig. 5.10. Current transient $J(t)$, showing the steady-state regime (flat) and the polarization relaxation regime (decreasing at short times t). Note that for impure samples the true steady-state leakage current is encountered at shorter times than in pure materials. This makes it easier to determine true leakage current in dirty specimens!

Two points must be made early in this discussion. First, the measurement of leakage current must be made in steady state, which requires that the sampling time be long compared with relaxation times and short compared to degradation (TDDB) times. As shown in Fig. 5.10, this will be sample-dependent. the leakage current data in some early papers in which long sampling times were not used were probably true steady state values, simply because of their large absolute values. Second, there is good reason to believe that most BST films have highly inhomogeneous trap densities near the electrode interface, with a large gradient in defect concentration. For example, Mihara et al. find 10^{18} cm^{-3} in the film interior but 5×10^{20} near the surface [191, 206]. At applied voltages the surface trap emptying process will necessarily give rise to space charge gradients and to band bending. This trap emptying model is used quantitatively by Dey et al. to explain their very recent depletion width $d(V)$ data [211].

5.1.2 Dependence of Schottky Barrier Height upon Electron Affinity

$$S = \frac{d\Phi_B}{d\chi} = 0.6 \, . \tag{5.8}$$

In an ionic dielectric which is fully depleted, the Schottky barrier height has a very simple dependence, discussed in Chap. 4, on the electron affinity of the semiconductor χ, or equivalently, upon the work function Φ of the metal electrode. As shown in Fig. 5.11, this gives a derivative of Φ with respect to χ (termed S in Sze's text [204] and elsewhere) which has slope unity and intercept equal to the ferroelectric semiconducting band gap. This value was thought by some [221, 222] to be 1.0 in oxide perovskites such as $SrTiO_3$, but two factors will render it less than unity. As first established [223] by Bardeen in 1947, covalent bonds [224] will make S much less than unity, and Robertson and Chen have pointed out [225] that the covalent Ti–O bonds in $SrTiO_3$ will make S approximately 0.3; and Schluter obtained an earlier value of about 0.5 [226]. In addition, the presence of traps (e.g., surface donors) in BST will al so give both an intercept much less than E_g and a slope much less than 1.0. Scott estimates $S = 0.7$ for strontium bismuth tantalate [209] due to traps.

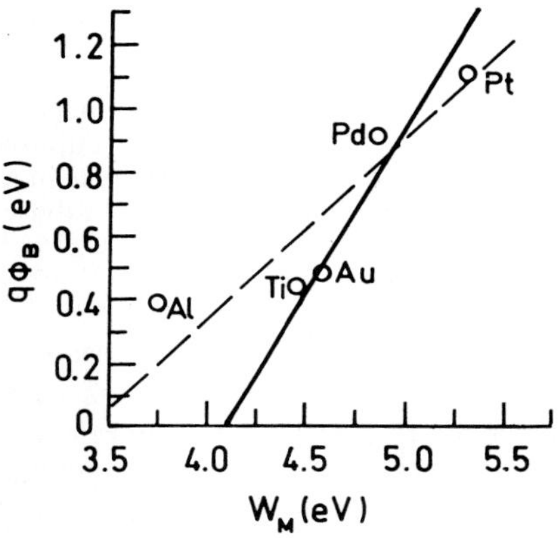

Fig. 5.11. Schottky barrier height Φ_B versus electrode work function Φ_M for BST on elemental metal electrodes. The slope gives the function S (the derivative of barrier height with respect to ferroelectric electron affinity χ) [202, 203, 209]

Thus in general the value $S = 0.5$ or so for BST militates against a simple trap-free ionic band matchup picture, such as that favoured by Brillson [221] or Kurtin [222], and reconciles the BST:Al electrode device data shown in Fig. 5.11. (Al electrodes yield a 0.4 eV Schottky barrier height experimentally, but the ionic trap-free model predicts ohmic contacts.) However, the degree of contribution of covalent Ti–O bonds and surface pinning states to this $S < 1$ result is not yet quantitatively clear, and further experiments on donor-doped and acceptor-doped BST will be required. These have been initiated by Wouters [227], by Dey [228], and by Klissurska [229] for PZT, but little has been reported on doped BST.

5.1.3 Dielectric Constant ε Appearing in Schottky Equation

When the standard Schottky model is used to describe leakage current data, the dielectric constant enters the equation (5.5). There is much discussion in the literature as to the proper value for this parameter. In a band tilting model (fully depleted film) the electrons must travel via either thermal hopping over the barrier or, at large fields, tunnel quantum mechanically through it (Fowler–Nordheim effect). In any case, the time it takes to get over or through a wide barrier (Fig. 5.1a) is not short. It can be estimated [209] using the electron drift velocity as long compared to the period of a long-wavelength optical phonon in $SrTiO_3$. In this case, the full dc dielectric constant $\varepsilon = 160$ should appear in (5.2). However, the experimental data in Fig. 5.12 [202] are compatible with the optical dielectric constant, which is simply the index of refraction squared $n^2 = 5.6$. This implies that the transit time through the Schottky barrier is much less than the period of an optical phonon. Using the known drift velocity for BST, Scott has inverted this argument to estimate the width of the Schottky barrier in BST (Fig. 5.1b) as 3 to 4 nm. For BST a similar result is obtained: $\varepsilon = n^2 = 4.8$ appears in the Schottky-fits to leakage current data $J(V)$ [202]. The earlier values of ε quoted in the literature, e.g., by Wang and Tseng [220] are incorrect, since they were based on the erroneous assumption of an electron mean free path $\lambda > d$, the film width.

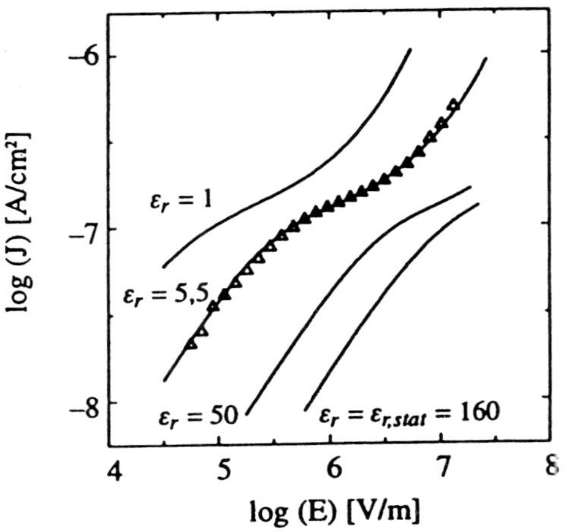

Fig. 5.12. Leakage current $J(E)$ versus applied field E [202, 203, 209]. The best fit yields a dielectric constant $\varepsilon = n^2 = 5.6$, where n is the index of refraction, and not equal to $\varepsilon_{dc} = 160$. This shows that the tunneling electrons through the interface barrier travel in a time fast compared to one optical phonon period, and hence that the tunneling width must be much less than d, the film width. Hence, the diagram in Fig. 5.1a cannot be correct if Fowler–Nordheim tunneling occurs, but this tunneling is confirmed experimentally in both [194] and [202]

Related to this argument is the observation of the onset of Fowler–Nordheim tunneling. A numerical solution by Dietz of an integral equation combining (5.6) and (5.7) yielded predicted theoretical regimes $J(E,T)$ for both thermionic Schottky-limited leakage current and true Fowler–Nordheim tunneling. Azuma and Scott initially measured [220] Fowler–Nordheim onset over the temperature range from 293–500 K and voltages from 0–12 V across 145 nm of BST, shown in Fig. 5.13. These measurements were extended by Dietz [202]. However, in all cases the observed onset is around 3.5×10^7 V/m at room temperature, increasing to approximately 6×10^7 V/m at 500 K. These values are an order of magnitude smaller than the theoretical calculations, as shown in Fig. 5.14. We believe that this arises simply from the incorrect assumption that the voltage drop is uniform across the full film thickness. If for example 50% of the drop were across a thin (5 or 6 nm) depletion width, then the field very near the electrode would be sufficiently high to produce Fowler–Nordheim tunneling, and the experimental data would agree with Dietz's theoretical calculation. Note, however, that Zafar et al. have successfully assumed $E = V/d$ in their BST/Pt films and derived a value of $\varepsilon = n^2$ compatible with that assumption.

A simpler qualitative argument is also useful here [230]. For a band tilting model with 5 V applied voltage across a 150 nm film and a Schottky barrier height of 1.0 V, the electrons in Fowler–Nordheim tunneling must tunnel through $(1/5) \times 150$ nm $= 30$ nm. This is most implausible. Tunneling through thin dielectric films to give significant leakage current values (many $\mathrm{nA/cm^2}$) requires much thinner barriers – of order 3–6 nm. Therefore a simple band-tilting model (Fig. 5.1b) is not plausible on the grounds that it prohibits Fowler–Nordheim tunneling, contrary to observations.

5.1.4 Schottky-Modified SCLC Theory

Simmons developed, in three separate papers, a rather detailed model of Schottky-modified space-charge-limited currents. This builds upon earlier work by Adirovich and Stoeckmann. The basic results are shown in Figs. 5.8 and 5.9. Figure 5.8 illustrates the resistance R of a thin Schottky barrier as a function of voltage. As is readily seen, R drops 10 or 12 orders of magnitude as V increases from 0 to 3 or 4 V. Because of the Frenkel–Poole scattering behind this surface barrier, this film will behave as a pair of resistors in series, but one of the two resistors is extraordinarily nonlinear. Thus at low voltages (0–2 V) the device behaves like a Schottky barrier to injected electrons. But at only slightly higher voltages (2–4 V) it is utterly transparent to these injected electrons, and the device resistance is instead dominated by Poole–Frenkel resistance in the film interior (as concluded, incidentally, for BST by Hwang et al. [192, 193, 194, 231]). Figure 5.9 shows this change even more clearly. Plotted here, in a separate paper from Simmons, are the leakage current magnitudes for both Fowler–Nordheim tunneling and Schottky thermionic emission. The parameters in both Figs. 5.8 and 5.9 are close to those for BST/Pt. We see that at about 1 or 2 V in barriers 3 to 6 nm wide the two leakage current mechanisms have the same 50% contribution; at higher voltages or thinner films, tunneling dominates. At lower voltages or thicker films, thermionic emission dominates. Simmons has not included the

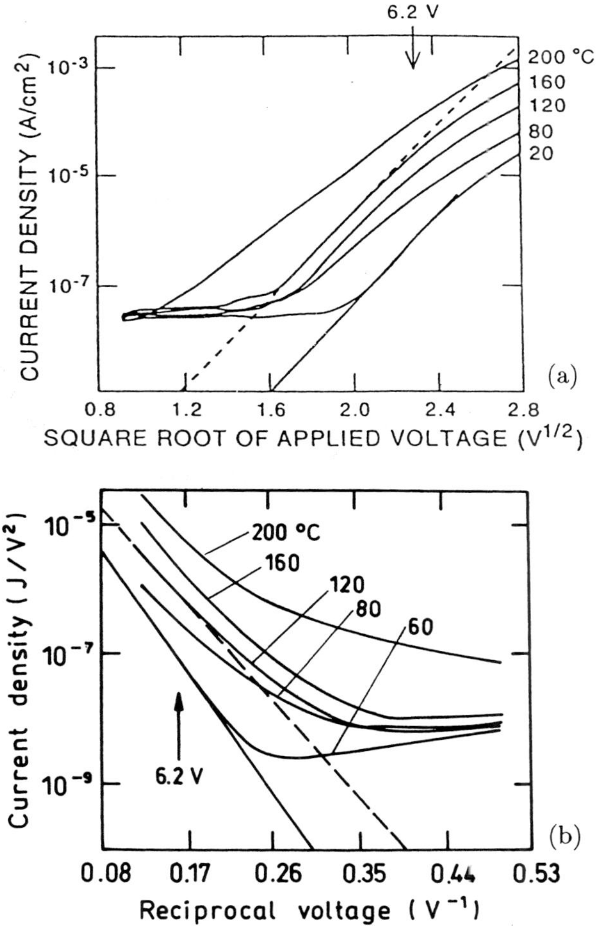

Fig. 5.13. (**a**) Leakage current versus $V^{1/2}$ in BST, showing Schottky behaviour up to about 8 V. (**b**) Leakage current divided by voltage squared versus $1/V$, showing Fowler–Nordheim regime in BST

Poole–Frenkel processes, but they will further limit the total current density in either case.

It is important to note explicitly that Simmons' complicated leakage current model is exactly linear in voltage at low V; that is, it will appear 'ohmic', as he discussed in detail in his paper. This can easily confuse experimental practitioners.

5.1.5 Space-Charge-Limited Currents

In all the cases discussed above, when the total leakage current reaches a certain threshold, independent of the mechanisms involved (which may be

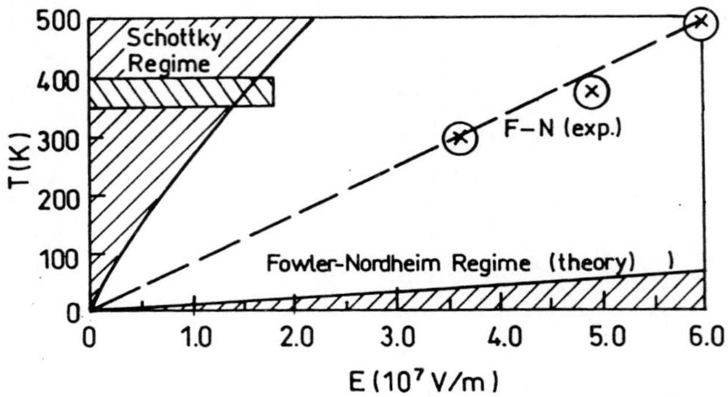

Fig. 5.14. Graph showing the regimes of Schottky and Fowler–Nordheim currents for BST/Pt. In this plot of temperature versus applied field the Fowler–Nordheim tunneling onset is at much lower fields [194] than predicted theoretically [202], suggesting that E is not equal to V/d, where d is the film thickness, but to V/d_d, where d_d is the depletion width

several), and independent of whether the electrode interface is ohmic or blocking, space charge will limit the current. This was first discussed by Adirovich [216] and by Stoeckmann [217]. The characteristics of SCLC are current magnitudes approximately quadratic in applied voltage (the exact exponent is discussed by Lampert [232]) and approximately cubic in inverse film thickness (the exact exponent is discussed by Mark and Lampert [232] and by Scott[233]).

There were several early reports of SCLC in PZT films, but only one ever of SCLC of electrons in BST (Krupanidhi). Note that SCLC of holes was observed as early as 1953 [234].

Both electronic and ionic SCLC exist in ferroelectric thin films. The ionic SCLC were well characterized by Zafar et al.

5.2 Arguments in Favor of a Completely Depleted Film

Against these arguments, which show that full depletion is reached only at about 5 V across 300 nm of PZT or BST, we have the observation by Outzourhit et al. that their thick strontium titanate films have $C(V)$ varying entirely due to the nonlinearity of the dielectric constant. Using an early (1961) theoretical model due to Johnson, they calculate

$$\frac{\varepsilon(T,\,E)}{\varepsilon(T,\,0)} = 0.64(1+x)^{-1/3}\,, \tag{5.9}$$

where

$$x = \text{constant}\ \varepsilon^3(T,\,0)\,E^2\,. \tag{5.10}$$

The experimental data in Fig. 5.15 fit this formula (5.9) very well.

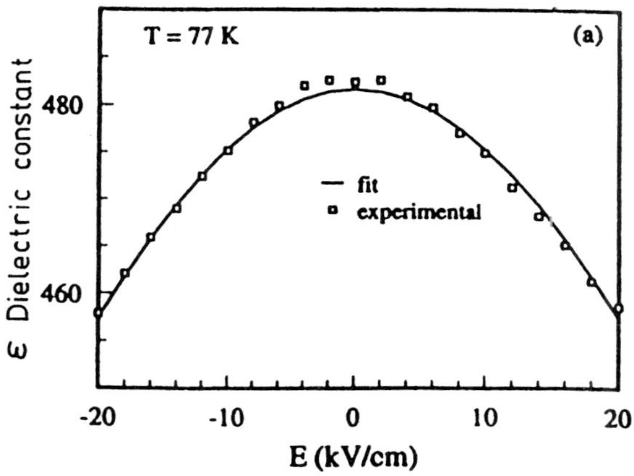

Fig. 5.15. $C(V)$ in thick strontium titanate films [194]

A more powerful argument is that impedance spectroscopy done both in Germany [202] and in Korea [193] on such films reveals no finite depletion width $< d$, the film thickness. We believe that the depletion width is too small to show up in the thick-film experiments of Outzourhit et al. (in which the field was applied interdigitally across the film surface width of 5 microns), and that the depletion width was too small to see via impedance spectroscopy; that is, depletion width was $\ll d$, not $\gg d$.

5.3 Zuleeg–Dey Model

There are many different expressions from theoretical models for the predicted dependence [224] of $C(V)$. We need to specify whether the space charge is assumed constant, independent of depth z into the film (pn-junction), or if it varies. The variation can be linear (graded junction) or quadratic (p-i-n junction). If we assume that the space charge varies with z, we must also consider whether the space charge width varies significantly with applied voltage V. In the case of assumption of constant space charge at low voltages (< 1 V), where the space charge density N and width d_d are both independent of V, $C(V)$ is given by

$$\frac{C}{A} = \left(\frac{q\varepsilon N}{V_\mathrm{bi} - V - k_\mathrm{B}T/e}\right)^{1/2} . \tag{5.11}$$

In this case the capacitance $C(V)$ varies as $V^{-1/2}$, or including a built-in bias voltage, as $(V_\mathrm{bi} - V)^{-1/2}$. This is the standard form expected for an abrupt pn-junction or for a Schottky barrier with a constant space charge density.

However, at higher voltages (2–5 V) it is necessary to worry about the change in space-charge width d_d with applied voltage V. Assuming that the interfacial capacitance of a Schottky barrier is given by [204]

$$\frac{C}{A} = \frac{\varepsilon}{d_\mathrm{d}} , \tag{5.12}$$

where A is the capacitor surface area, d_d is the depletion width, and $d_\mathrm{d} \ll d$, the film thickness; and postulating a quadratic dependence of space charge N near the surface

$$N(z) = N_0 \frac{z_0^2}{z^2} , \tag{5.13}$$

where N_0 is the charge concentration at a distance z_0 very near the surface, Dey et al. [228] derive for low voltages that

$$\log \frac{A}{C} = \varepsilon \frac{V_\mathrm{bi} - V}{q N_0 z_0^2} - \log \frac{\varepsilon}{z_0} , \tag{5.14}$$

where V_bi is the built-in bias voltage (due perhaps to charged defects). This formula yields $\log C$ linear in V and was first shown to fit BST data by Scott et al. [210], as shown in Fig. 5.16, and later shown to fit PLT (lead lanthanum titanate) [228], as illustrated in Fig. 5.17.

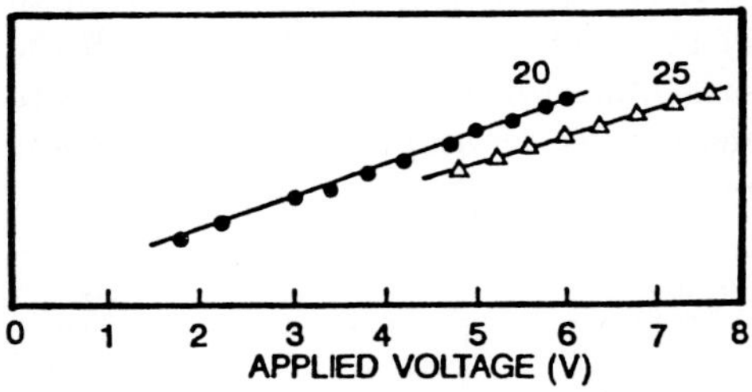

Fig. 5.16. $C(V)$ in BST films, showing agreement with (5.14) [194]. Log C is plotted versus voltage

The assumption that the charge density $\rho(z)$ varies quadratically with depth z into the film is plausible but not certain. If we make a different assumption, say that of a linearly graded junction, we obtain the result that $C(V)$ varies as $V^{-1/3}$ (see [204], p.82). In general we find

$$C = A \left(\frac{\varepsilon_\mathrm{s}}{\varepsilon N}\right)^{1/2} (V_\mathrm{bi} - V)^{-1/n} , \tag{5.15}$$

where ε_s is the dielectric constant for the surface layer and ε that of the film interior with n differing from 2 in (5.11). The situation with $n = 4$ can be observed in ferroelectric thin films, as illustrated in Fig. 5.18 for PZT/Pt from Symetrix.

In general, $C(V)$ can therefore be quartic, cubic, quadratic, or logarithmic, depending upon the voltage range and the space charge distribution $\rho(z)$ assumed as a function of depth z into the film; and all of these dependences have been reported experimentally. As a result, some authors [191] simply use

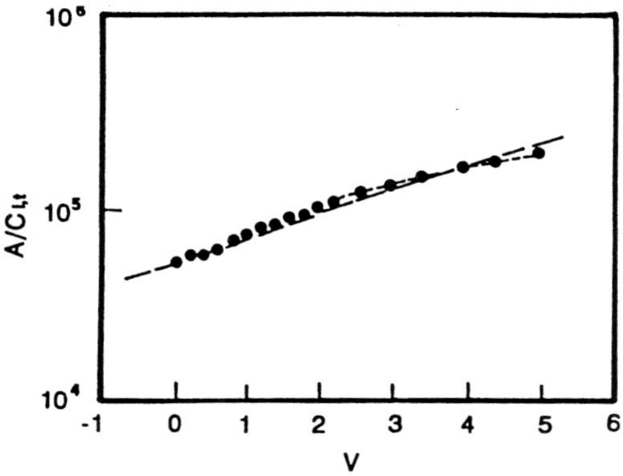

Fig. 5.17. $C(V)$ in PLT films, showing agreement with (5.14) [228]

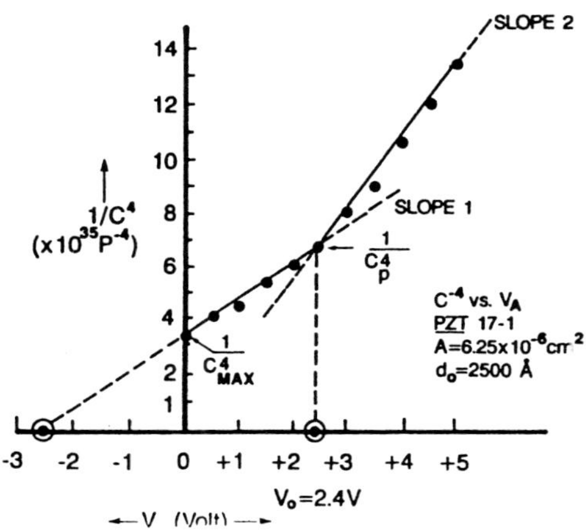

Fig. 5.18. $C(V)$ in PZT films, showing agreement with a different model, that of (5.15) [201]

n as an empirical fitting parameter; and indeed [191] find $n = 3$ for some PZT specimens, exactly in agreement with that for a linearly graded junction. As Lampert and Mark point out [205], for space charge injection in the trap-free limit, the charge distribution

$$N(z) = \text{constant } z^{-1/2} \tag{5.16}$$

is expected. However, as one gets closer to the ohmic limit,

$$N(z) = \text{constant } \exp(1 - z/d) , \tag{5.17}$$

so that a continuous range of effective exponents n is possible for real ferroelectric films.

5.4 Combined Model

At present the exact nature of the electrode–ferroelectric interfacial barriers in PZT, BST, and SBT is under discussion and unresolved. The present author favors the combined 'Matterhorn' model of Fig. 5.1b, in which both Schottky thermionic emission and Fowler–Nordheim quantum mechanical tunneling occur, the latter through only a 'tip' of approximately 5 nm wide. This question can be resolved in the near future by careful measurements of leakage current J as a function of temperature (Fowler–Nordheim tunneling has only a small dependence upon T at cryogenic temperatures; Schottky thermionic currents have a large T-dependence.) Such measurements have been initiated this year in Lausanne [235]. Experiments should also emphasize studies as functions of film thickness. A big field over a thick film gives the same result as a small field over a thin film in the band-tilting models; it does not in the Matterhorn model. At present the most convincing data are those [211] of Dey (Fig. 5.19).

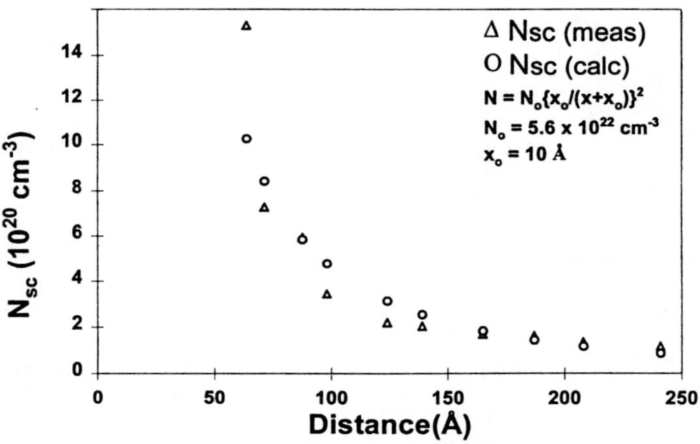

Bias, V	W_d, Å
0	63.9
0.5	71.5
1.0	87.8
1.5	98.1
2.0	124
2.5	139
3.0	165
3.5	187
4.0	208
4.5	241
8.0	377

Fig. 5.19. Space charge density $N(z)$ in strontium titanate, showing depletion with varying from 6.4 nm at $V = 0$ to 38 nm at 8.0 V [211]

It is useful in any model of depletion widths to make distinctions between films that are ferroelectric but below saturation voltage, films that are ferroelectric and above saturation voltage, and films that are paraelectric [236]. Only in the case of ferroelectrics above saturation voltage does it appear likely that single-carrier space-charge-limited currents can prevail in high-resistivity ferroelectric thin films.

5.5 Relationship to Band Structure Matchups Based upon XPS

All of these models have a very important effect on our understanding of the band matchups of ferroelectrics on metal electrodes. The key term is the image field reduction term. For BST DaCruz and her group at Symetrix [236] and the UNSW group of Scott [237, 238] have independently estimated a very large image field reduction term of about 0.8 eV at 3 V applied to BST films. This reduction term is quite comparable to the Schottky barrier height itself (of order 1 eV), so it introduces a 40% correction, not a small perturbation.

The importance of the present work is that it shows that we are unsure of the field E that goes into the image field reduction term, which is of form

$$\Delta\Phi_{\mathrm{B}} = \left(\frac{eE}{4\pi\varepsilon}\right)^{1/2} , \tag{5.18}$$

But it is not clear that $E = V/d$ here; nor is it certain what value of ε applies.

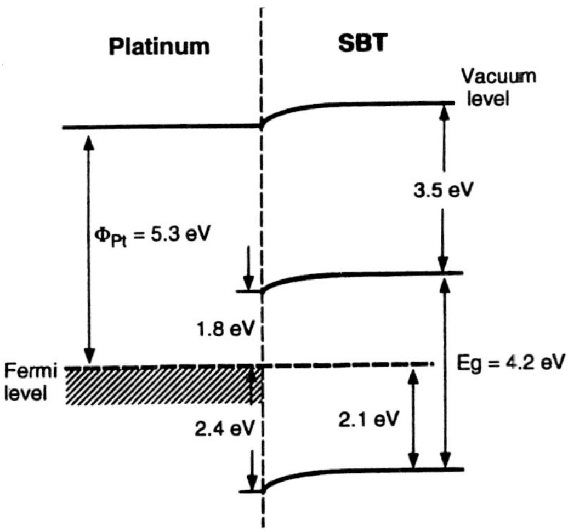

Fig. 5.20. Band structure in SBT/Pt

Table 5.1 gives numerical values for band matchups in BST, PZT, and SBT on Pt, Bi, RuO_2, and $SrRuO_3$ electrodes. The energy levels versus depth from the electrode interface are shown for SBT/Pt in Fig. 5.20 and PZT/Pt in Fig. 5.21. Figure 5.22 shows luminescence data for PZT films that may confirm the 1.8 and 2.3 eV energies estimated in Fig. 5.21 for PZT. Note that the band-bending is upward in our PZT specimens and the conduction is n-type. It is also useful to point out that even in PZT specimens in which the Fermi level is nearly mid-gap (compensated), the conduction can be predominantly n-type; this is because the electron mobility is about twice that

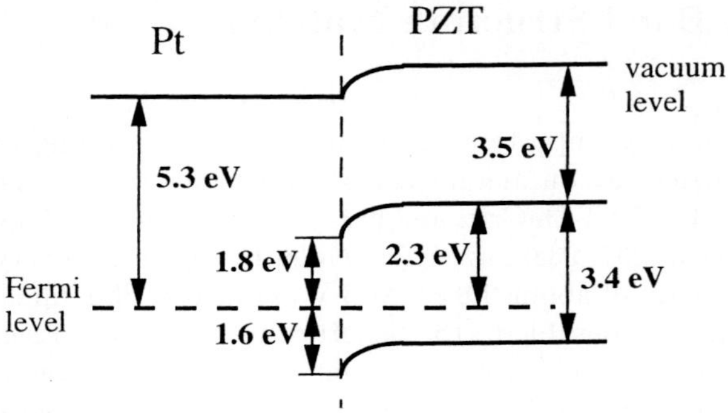

Fig. 5.21. Band structure in PZT/Pt

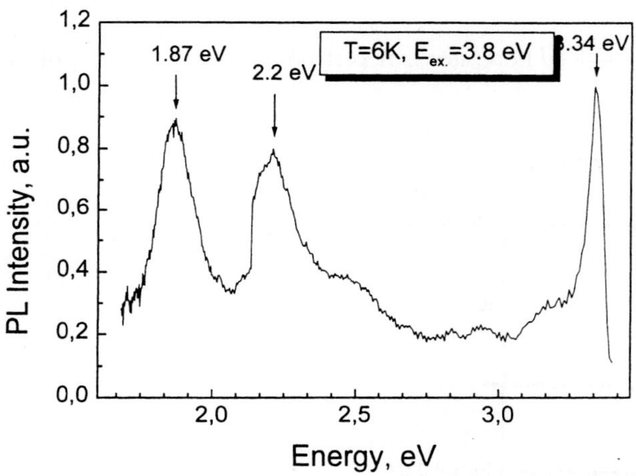

Fig. 5.22. Possible confirmation of these states by luminescence (Kovalev D., Tech. Univ. Munich, private communication 1998)

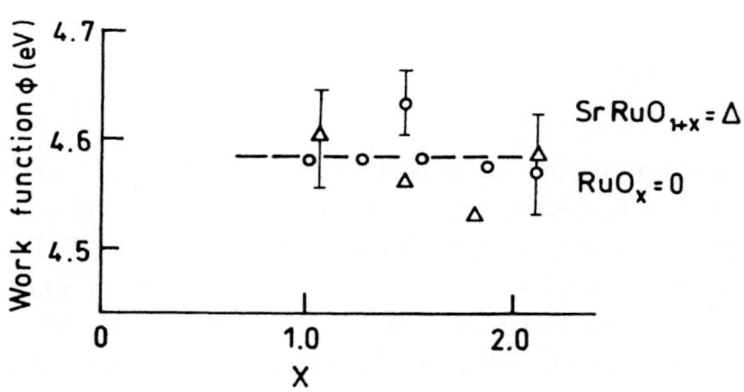

Fig. 5.23. Work function in $SrRuO_{1+x}$ and RuO_x versus x

Table 5.1. Band parameters for BST, SBT and PZT on Pt metal electrodes [eV]: $\Phi_M(Pt) = 5.3$ eV; $\Phi_M(RuO_2) = 4.8$ eV $= \Phi_M(SrRuO_3)$

Material	SBT	BST	PZT
Electron affinity χ	3.4 ± 0.2	4.1 ± 0.2	3.5 ± 0.2
Bandgap E_g	4.0 ± 0.1	3.2 ± 0.1	3.4 ± 0.1
Fermi level pinning energy Φ_0 above valence band (sample-dependent)	2.1 ± 0.2	1.8 ± 0.2	1.0 ± 0.2
Band bending at $V = 0$ in FE	0.3 ± 0.1	0.2 ± 0.1	0.5 ± 0.1
Schottky barrier height Φ_B at $V = 0$ (calculated from Φ_M, χ)	1.8 ± 0.2	1.5 ± 0.2[a] 1.60 ± 0.1[c] 1.63 ± 0.1[d]	1.8 ± 0.2[b]
Schottky barrier height Φ_B at 3 V (calculated)	0.9 ± 0.2	0.6 ± 0.2	1.0 ± 0.1
Image field reduction at 3 V	0.8 ± 0.2	0.8 ± 0.1[a]	0.8 ± 0.1
Schottky barrier height at 3 V from experimental $J(V)$	0.9 ± 0.1[f]	0.7 ± 0.2[a] 1.1 ± 0.2[g] 0.6 ± 0.1[j]	1.5 ± 0.2[e] 1.0 ± 0.2[h]

[a] Joshi et al., Integ. Ferroelec. **14**, 133 (1997).
[b] Dey et al., Jpn. J. Appl. Phys. **34**, 3134 (1995).
[c] Scott et al., Integ. Ferroelec. **4**, 61 (1994).
[d] Hwang et al., J. Appl. Phys. **83**, 3703 (1998).
[e] Scott, Jpn. J. Appl. Phys. **38**, 2272 (1999).
[f] Waser, Sci. & Tech. Electroceram. (Kluwer, Dordrecht, 1995) p.223.
[g] Watanabe et al., Integ. Ferroelec. **22**, 241 (1998); **17**, 451 (1997).
[h] Stolichnov and Tagantsev, J. Appl. Phys. **84**, 3216 (1998).
[j] Zafar et al., Proc. ISAF (Lausanne, Sept. 1998, in press); Appl. Phys. Lett. **73**, 1 (1998).

for holes, so given an equal number of thermally excited electrons and holes, the electron current will be larger. (Note that although the electron mobility is around 0.2 cm^2/Vs in pure strontium titanate, it is 1.0×10^{-3} cm^2/Vs in BST thin films [239].) The rather ohmic result for BST films on RuO_2 or $SrRuO_3$ electrodes agrees with the comment from LG Institute of Technology that due "to their high leakage currents (they are) not applicable to practical DRAM applications" [239]. The kinetics of oxidation of Ru to produce RuO_2 are interesting in this regard. RTP of ruthenium is discussed by Tanaka and Ami [239]. We have found recently that the metal work function for RuO_2 and RuO_{2-x} is 4.6 eV, and at first glance appears to be independent of oxygen percentage x (including elemental Ru metal) (Fig. 5.23). The same value of 4.6 eV appears to be the work function for $SrRuO_3$. However, what newest XPS data show is that the oxidation of an Ru electrode surface is inhomogeneous. Below 600°C islands of metallic Ru coexist with RuO_2 or $SrRuO_3$. The Ru^{+4} in RuO_2 or $SrRuO_3$ produces a work function of 4.95 eV and a large Schottky barrier height. The metallic Ru produces a work function of 4.6 eV and a smaller Schottky barrier. Thus leakage current in a BST/Ru device will be spatially inhomogeneous, leading to 'hot spots' on the film surface; the current will be controlled by the metallic Ru spots. This also leads to dendritic breakdown, as shown by Raleigh, in oxide electroded dielectrics. Process engineers can eliminate this problem by annealing the Ru above 600°C, a temperature we find sufficient to produce uniform oxidation of Ru to RuO_2 (or $SrRuO_3$). Thus, BST/Ru films processed at 500°C will have Ru work functions of 4.6 eV and smaller Schottky barrier heights (higher leakage currents) than BST/Ru films processed at 620°C, which have work functions of 4.95 eV.

RuO_2 is unusual in that Ru retains its $4d$ electrons when it crystallizes in the rutile structure, with the result that its Fermi surface lies in the Ru d-states. Hence it is almost completely like elemental Ru. Thus the change in work function with oxidation state of the Ru might be expected to be small. The band structure and density of states were calculated by Chelikowski and Glassford [239]. However, the change in work function upon oxidation is not much different from that of Pt. We note in Table 5.1 that the most recent experimental values for Schottky barrier heights of 0.65 ± 0.06 eV for BST from Zafar et al. and 1.0 eV for PZT from Stolichnov and Tagantsev agree exactly with our calculations from band structure matchups (Φ_M andχ). The SBT barrier height has also been confirmed independently, most recently by Seong and Yoon [240]. We note parenthetically that the t_{2g} band in RuO_2 is 1 eV higher than in IrO_2, indicating the greater oxygen affinity of Ru [239]. This may be very significant in choosing between Ru and Ir electrodes and in quantitative choices of processing parameters.

5.6 Ionic Space-Charge-Limited Currents

It was pointed out in 1994 by the author [220] that space-charge-limited currents in ferroelectric films in general exhibit time dependences that are not monotonically decreasing. A linear rise to a cusp is usually observed, followed by exponential decay. Near $t = 0$ the initial leakage current $J(0)$ is given by

$$\varepsilon = 2.25 \times 10^{13} J(0) \frac{d^2}{\mu V^2} \,. \tag{5.19}$$

For BST films the numerical values of J measured by the present author gave an estimate of $\mu = 10^{-7}$ cm^2/V s, and it was noted that this was close to the oxygen vacancy mobility determined by Waser, Baiatu et al. (By comparison, hole mobility is about 0.1 cm^2/Vs and electron mobility 0.2 in pure $SrTiO_3$; 10^{-3} cm^2/Vs in BST *films* is presumably trap-limited. This much lower mobility – a 100 times less – in BST films is due to the greatly increased film trap density [241], and we note that Mihara [191, 206] finds interface trap densities about a 100 times greater in PZT or BST films compared with bulk.) However, this value predicted a time t (max) at which leakage current would peak in PZT or BST at room temperature which is too short for most measurements. Perhaps more important, Many showed that no cusp will be observed if the trapping time is shorter than the transit time. Thus in PZT in might be necessary to make measurements on extremely thin films to see this effect.

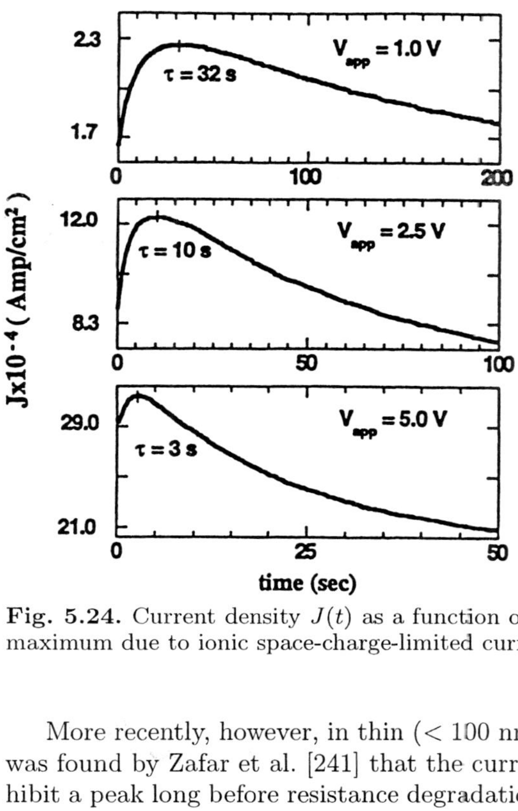

Fig. 5.24. Current density $J(t)$ as a function of time t in BST films, showing the maximum due to ionic space-charge-limited currents [240]

More recently, however, in thin (< 100 nm) BST films on Pt electrodes it was found by Zafar et al. [241] that the current density $J(t)$ does indeed exhibit a peak long before resistance degradation sets it. As shown in Fig. 5.24, this peak is at a few seconds, depending upon exact applied voltage, for voltages from 1 to 5 V. The exact peak position is predicted by Many and Rakavy [241] to occur at a time

$$t\,(\max) = \frac{0.78d^2}{\mu V}\,, \tag{5.20}$$

where d is the film thickness, μ the carrier mobility, and V applied voltage. The peaks were observed to occur not in the [ms] regime, as estimated by Scott et al. [220], but in the 1–50 s regime. This is due to the fact that the actual ionic mobilities were found to be of order 10^{-12}, not the 10^{-7} estimated by Scott. These peaks may occur for space-charge-limited currents of electrons in ferroelectrics also, and not just ions, but in that case (5.20) predicts a peak in the [ns] regime, too fast for most experimenters to see. However, if μ is the ionic mobility (about 2×10^{-12} cm^2/V s in BST), the peak after a few seconds is easily seen (Fig. 5.24). We note that *ionic* space-charge-limited transient currents have been known in SiO_2 for more than thirty years [241], and also studied carefully in silicon by Lemke and Mueller [241]. Therefore their occurrence in BST is not wholly unexpected or controversial.

The oxygen vacancy mobility is estimated independently from the Einstein relationship

$$k_B T \mu = qD \,, \tag{5.21}$$

where D is the diffusivity. Oxygen vacancy diffusion is known to obey

$$D = g \exp\left(-\frac{E_{ox}}{k_B T}\right) , \tag{5.22}$$

where E_{ox} is the activation energy for oxygen diffusion (found to be 1.0–1.1 eV in agreement with independent values).

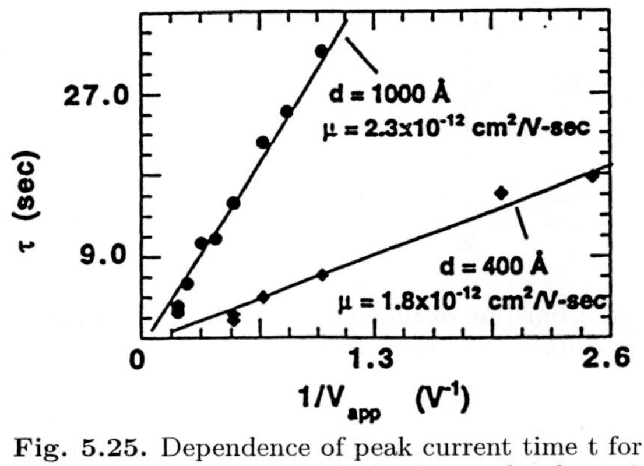

Fig. 5.25. Dependence of peak current time t for BST capacitors 40 and 100 nm thick upon reciprocal applied voltage; the slope gives the oxygen vacancy mobility [240]

Figure 5.25 shows a plot of peak time t in (5.20) versus $1/V$. The slopes yield oxygen vacancy mobilities of 2.3×10^{-12} cm^2/V s for a 100 nm film and 1.8×10^{-12} cm^2/V s for a 40 nm film. Figure 5.26 shows the dependence of the measured mobilities for the two thicknesses versus reciprocal temperatures. The observed slopes correspond to 1.0–1.1 eV activation energies.

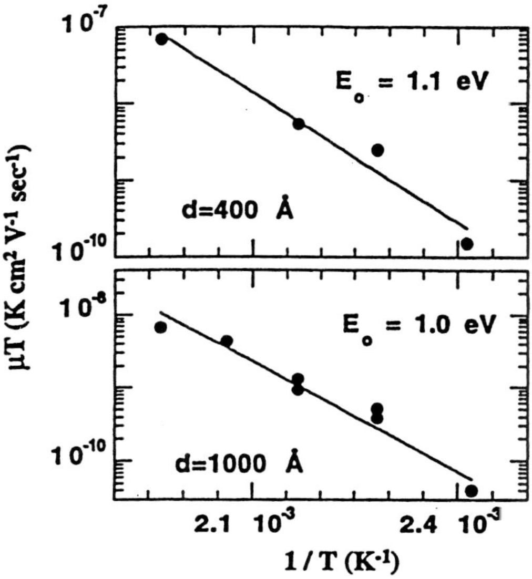

Fig. 5.26. Dependence of μT upon reciprocal temperature for BST films, giving the activation energy E_{ox} for oxygen vacancy conduction [240]

This work not only nicely confirms the earlier predictions of Many [241], it also shows unambiguously that space-charge-limited currents do indeed occur in ferroelectric thin films, particularly BST. There had been controversy about that in the case of electron conduction; but for ionic conduction transients, the case is quite clear due to the observation of a maximum in the $J(t)$ curves. Parenthetically, we note that small polaron models of oxygen vacancy conduction cannot be used in ceramics such as fine-grained BST because of the effects of the grain boundaries [242]. As a result, much modeling has been carried out for these systems recently [243]. Modeling each grain as a resistor with a parallel capacitor, one finds that there is a characteristic dispersion frequency of around 100 kHz if grain boundaries dominate resistance, whereas the characteristic dispersion frequency is several orders of magnitude higher (many MHz) at the polaron hopping frequency, if grain boundary effects are negligible [244, 245, 246]. Studies of strontium titanate in the 1960s failed to reveal any evidence for dc hopping of small polarons in strontium titanate, but the topic is still debatable and unresolved [247, 248].

Finally we note that the mere observation of space-charge-limited currents tells us that there is a reservoir of the carrier types being injected. The current level sustainable in a long pulse gives information about the chemical potential, in the case of oxygen ion space-charge-limited currents, and about the contact potential, in the case of electron space-charge-limited currents. Smith and Rose inferred [249] that the Fermi level in In is < 0.3 eV above the CdS conduction band from the fact that a space-charge-limited current of

20 A/cm^2 could be produced. These ideas suggest ways of further exploiting the ionic space-charge-limited transients discovered by Zafar et al. In particular, her BST work should be repeated with oxide electrodes replacing Pt.

6. Switching Kinetics

In principle when a large positive electric field is applied to a ferroelectric material that is in a multidomain state with most domains antiparallel to the applied field (a negative polarization state) the negative domains could all reorient simultaneously and independently. In fact, within Landau theory [250] one can calculate exactly what coercive field would be required in this case. However, the numerical result is typically approximately 1000 times larger than the coercive field observed experimentally. This shows that in practically all cases it is energetically cheaper to switch not by simultaneous polarization within a domain but rather by moving the domain walls. With one exception – the switching of ultrathin 0.9 nm films of polyvinylidene-tetrafluoro-ethylene copolymer Langmuir–Blodgett films [251], which appears to be true Landau theory simultaneous polarization flipping – all ferroelectrics at achievable field strengths E appear to switch by moving domain walls.

The switching process most commonly encountered proceeds as shown in Fig. 6.1. First, nucleation occurs, primarily as inhomogeneous nucleation at surfaces. This process involves fluctuations and therefore requires a certain

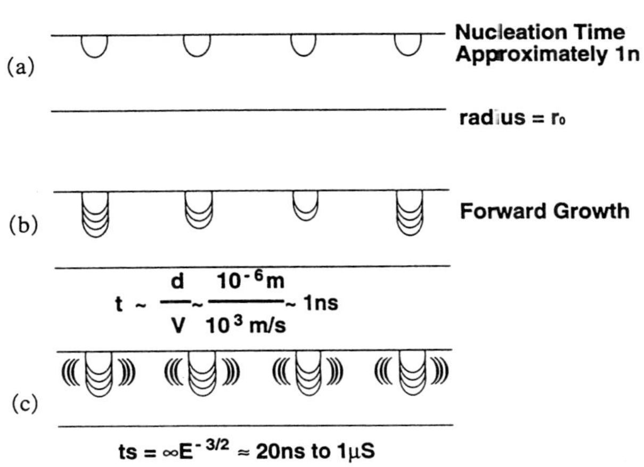

Fig. 6.1. Schematic diagram of switching in a ferroelectric film. (**a**) Inhomogeneous nucleation at surface. (**b**) Forward growth of needle-like domains parallel and antiparallel to the applied field. (**c**) Sideways growth of domains. In reality these processes may occur simultaneously

time t_n. The physics is analogous to the bursting of a soap bubble. In that case a microscopic void arises on the bubble surface. However, the void does not burst the bubble (it is too small). Rather, it fluctuates in size, sometimes disappearing, sometimes growing. Only when it reaches a certain critical size r_c does it begin to grow exponentially in time. This bursts the bubble. Similarly, a positive polarity domain that nucleates on the surface of a ferroelectric that had been polarized negatively will fluctuate in size (sometimes disappearing and reappearing). Only when it reaches a critical size does it begin to grow quickly. This nucleation time is typically 1 ns in oxide ferroelectrics, and the critical size is typically about 1–10 nm. After that the domain spreads out rapidly in a needle-like geometry, growing parallel or antiparallel to the applied electric field. Unlike magnetic domain walls, ferroelectric domain walls do not seem to move supersonically. Thus the time for a domain to propagate from the anode to cathode in a capacitor film may be estimated as

$$t_{fg} = \frac{d}{v} , \tag{6.1}$$

where v is the speed of sound. This regime is called the *forward growth* regime, and for a film a few microns thick with a sound velocity a few [km/s], this is also approximately 1 ns, like the nucleation time. Thus, before the domains have had time to spread out sideways and fill up the entire film volume, we have about 2 ns in the switching process. This suggests that in the average ferroelectric, sub-nanosecond switching is unlikely.

Just because the needle-like domains have spanned the film from anode to cathode in 2 ns does not mean that a memory device using polarization can be read in 2 ns. Perhaps as little as 1% of the total film has reversed polarization, which is what the sense amplifiers of the RAM must read as displacement current. In order for the remaining 99% to switch, the domains must spread out sideways to fill the entire film. This sideways growth time t_s can be slow – a few [ns] to a [μs] – and typically it varies as applied field

$$t_s = \text{constant } E^{-3/2} . \tag{6.2}$$

The relative values of nucleation time, forward growth time, and sideways growth time differ in different ferroelectric materials. For example, in Guanidinium Aluminium Sulfate Hexahydrate (GASH) the nucleation time is the slowest, rate-limiting parameter. In ABO_3 perovskites such as $BaTiO_3$, the sideways growth time is slowest (and hence rate-limiting) in bulk or in thick films, but not necessarily in thin films, as discussed below. In bulk TriGlycine Sulfate (TGS), they are about the same. This in turn results in quite different graphs of displacement current $i(t)$ versus time t after the external field is applied. For perovskites or Aurivillius phase ferroelectric oxides, the resulting $i(t)$ curves are symmetric and rather Gaussian in appearance, as shown in Fig. 6.2, but one should not assume that this will be true in other ferroelectric film materials.

Switching kinetics in ferroelectric thin films is not as simple as most textbooks would suggest. One complication is that the coercive field E_c is not a

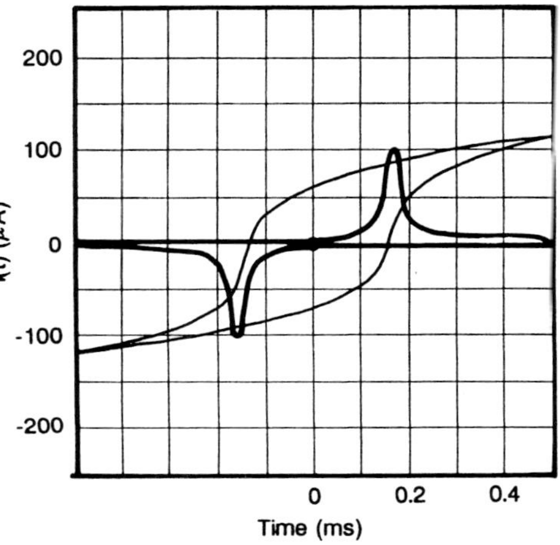

Fig. 6.2. Displacement current $i(t)$ versus time t after the external field is applied in PZT. Note that this is only displacement current $i(t) = \mathrm{d}P(t)/\mathrm{d}t$. The ohmic conduction is negligible in these highly insulating films at low voltages

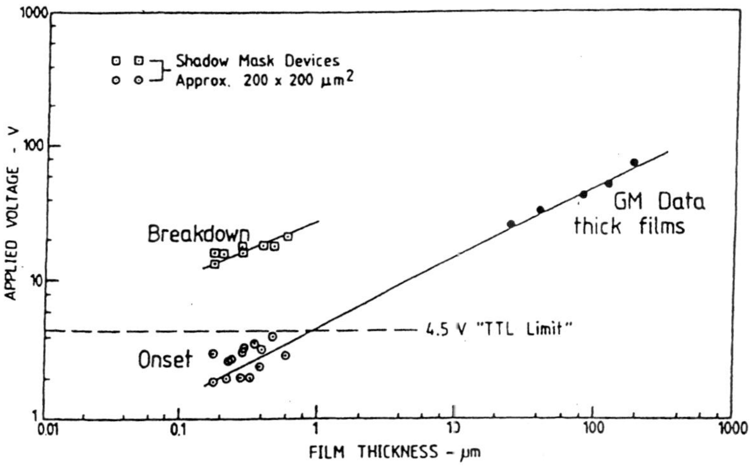

Fig. 6.3. Coercive voltage V_c versus film thickness d for KNO_3 films [251]

constant but instead depends strongly upon film thickness. For nearly forty years the theory of Kay and Dunn [252] has been successful in describing switching in thick ferroelectric films and in particular the dependence of coercive field E_c upon film thickness d. However, their model is explicitly a low-field approximation suitable for thick films. Hence, for the very thin ferroelectric films in use for nonvolatile RAMs, it is expected to fail. Figure 6.3 shows that it works [253] perfectly well for KNO_3 films down to 200 nm, however, and Fig. 6.4 shows even more surprisingly that it is valid [254] in PZT

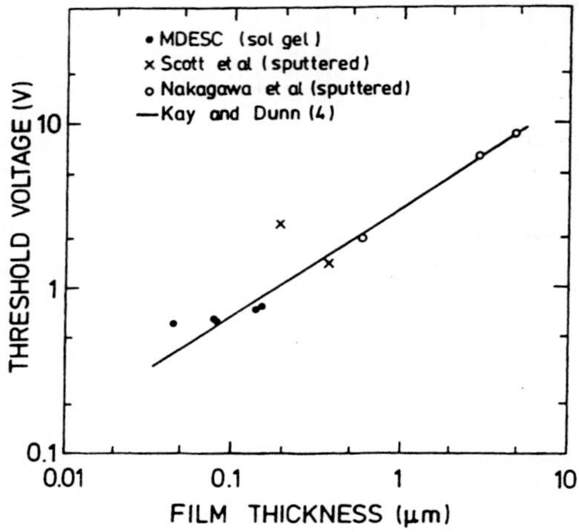

Fig. 6.4. Coercive voltage V_c versus film thickness d for PZT films [252]

films down to 40 nm. Other studies of the dependence $E_c(d)$ in ferroelectric thin films find similar results [255, 256].

Typically the nucleation problem in ferroelectrics has been treated in two approximations. Either all the nucleation is inhomogeneous (mostly at surfaces) as described above, and domain walls move from cathode to anode (or vice versa); or nucleation is homogeneous throughout the film interior and occurs at random sites. In fact, as shown in Fig. 6.5 [257], nucleation can occur coherently as microdomains in front of a moving domain wall. This occurs because ferroelectric domain walls are in general charged and hence exert a Coulombic field as they move through the dielectric [258].

Typically in ferroelectric switching the displacement current pulse $i(t)$ is detected in a time short compared with the transit time of a sound wave. This implies therefore that either the domains are moving supersonically or that there is substantial nucleation in the interior of the film (either homogeneous nucleation or, more likely, inhomogeneous nucleation at defect sites). It is

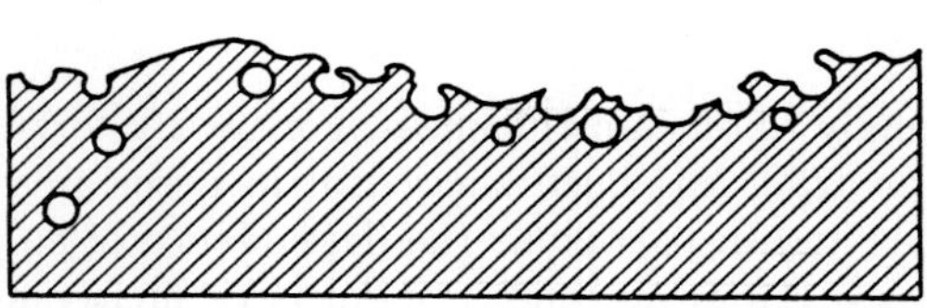

Fig. 6.5. Optical micrograph of microdomains nucleating in front of an existing domain wall in ferroelectric lead germanate. Scale is approximately 1 μm/cm [257]

my belief that ferroelectric domain walls are never supersonic. Evidence for this comes from comparing behaviour of ferroelectric and magnetic domains. With large applied magnetic fields B it is quite possible to drive magnetic domain walls at supersonic speeds [259]. But when this is done, acoustic phonons are emitted at a phase-matched angle like bow waves on a ship. This Cerenkov-like emission is never observed in ferroelectric switching. I believe that the difference in the ferromagnetic and ferroelectric cases is that ferroelectric domain walls are charged, whereas magnetic domain walls are not. In the ferroelectric case, the coherent nucleation of microdomains in front of an advancing macroscopic domain illustrated in Fig. 6 5 implies that a snow-plough effect occurs, such that the group velocity of the advancing domain front can be supersonic while the phase velocity of an individual unit cell polarization is subsonic.

Fig. 6.6. Labyrinthine domains on the surface of lead germanate. Scale is 1 μm (*marker*) [254]

A final complication, illustrated in Fig. 6.6 (marker scale is one micron), is that ferroelectric domains need not always be rectilinear. They can be spaghetti-like, often called *labyrinthine*, particularly at surfaces [260, 261].

Of course the experimentally measured switching times may be limited by more pedestrian causes, such as the RC-time constant of the measuring apparatus or the time required for the voltage source to deliver sufficient current to the capacitor. The latter depends critically on the surface area A of the capacitor, and so it is useful to make measurements on as small capacitors as possible. The former depends critically on the load resistance over which the displacement current is measured as a voltage drop. Early work (ca. 1985) often used 10 kΩ resistors, which badly limited switching speed measurements. Figure 6.7 illustrates the effect of reducing cell area from 4×10^4 μm^2 to 120 μm^2 and load resistance from 10 kΩ to 10 Ω. The effect of doing both was to reduce measured switching time from 1 ms to 60 ns in KNO_3 ferroelectric film capacitors. In the latter high-speed regime, the variation of switching time t with applied electric field E varies as in (6.2). But for larger resistive loads R_L or larger capacitor areas A, it does not, varying instead as

$$t = \frac{Q}{i} = \frac{2P_r A R_L d}{E} = \text{constant}\ \frac{1}{E}\,, \tag{6.3}$$

as shown in Fig. 6.7.

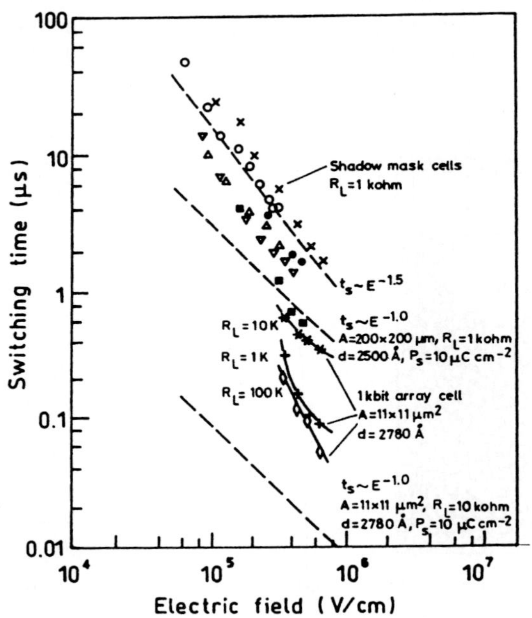

Fig. 6.7. Switching time versus applied electric field (**a**) in KNO_3 films [253], (**b**) in PZT

It is useful for device construction to make numerical estimates to determine whether in small area capacitors the nucleation time or the sideways growth time will be more important. The domain-speed limited switching time is given theoretically [262] by

$$t_s = \left(\frac{2N_n R v^2}{9}\right)^{-1/3}, \tag{6.4}$$

where N_n is the number of nucleation sites per unit area, v the domain wall speed, and R the nucleation rate at a particular applied field E. The nucleation time is given by

$$t_n = (N_n R A)^{-1}\,, \tag{6.5}$$

where A is capacitor cell area. Hence for nucleation time to be negligible, we require

$$N_n RA \gg \left(N_n R v^2\right)^{1/3} , \tag{6.6}$$

which is

$$\frac{v}{N_n R} \ll A^{3/2} . \tag{6.7}$$

Theoretically both v and R vary [262] as $E^{3/2}$, so (6.7) is independent of E. Using values of R measured by fitting the shape of displacement current transients $i(t)$ it is found that nucleation time t_n will dominate total switching time in ferroelectric memory cells smaller than 13 μm on a side for KNO_3. For PZT where $R = 1.5 \times 10^{22}\ \mathrm{s^{-1}cm^{-2}}$ [263], nucleation time becomes dominant for cells 1 μm^2 or smaller. Physically, this just means that in very small cells the time it takes a domain wall to propagate to the edge of the cell becomes short compared to the nucleation time, whereas in larger cells of the same material, the domain wall propagation time is long compared to the nucleation time. For practical memory devices at the Mbit–Gbit scale, it means that domain wall motion is unimportant; only nucleation time is important. Ironically, ferroelectric cells with high surface defect density will therefore have the fas test switching times.

There have been two popular theoretical models for domain switching times. The first [262] assumes, as outlined above, that there are three basic parameters – nucleation time, forward growth time, and sideways growth time. In this theory the expression for displacement current is

$$i(t) = \left[\frac{t}{\tau_R} - 1 - \exp\left(-\frac{t}{\tau_R}\right)\right] \exp\left\{-G\left[1 - \frac{t}{\tau_R} - \frac{0.5t^2}{\tau_R^2} - \exp\left(-\frac{t}{\tau_R}\right)\right]\right\} , \tag{6.8}$$

where τ_R is a relaxation time for domain motion (related to t_s) and G is a dimensionless parameter related to critical domain nucleation size. In KNO_3 thin films $P_s = 15\ \mu\mathrm{C/cm^2}$, $\tau_R = 120$ ns, and $G = 10.5$ (corresponding to critical domain radius $r_c = 1$ nm), as shown in Fig. 6.8.

The second switching theory [264] also parameterizes the equations with three constants, but these are domain speed, domain dimension D ($D = 1$, 2 or 3), and nucleation time. It is possible to relate these three fitting parameters [265] to those in the first theory. In this second model the displacement current is given by

$$i(t) = \frac{2P_s AD}{t_0}\left(\frac{t}{t_0}\right)^{D-1} \exp\left[-\left(\frac{t}{t_0}\right)^D\right] , \tag{6.9}$$

where t_0 is an effective switching time related to both nucleation time t_n and domain wall speed v.

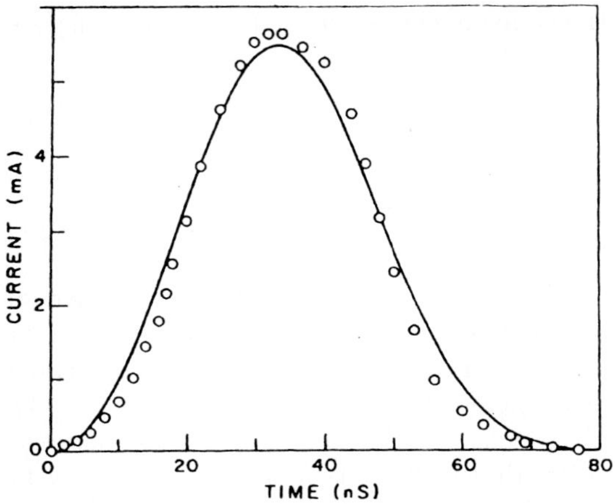

Fig. 6.8. $i(t)$ in KNO_3 fitted to (6.8)

When experimental data are fitted to this formula, two parameters are of importance. First, the average domain dimension D, which will be non-integer because the statistics average over needle-like ($D = 1$), pancake-like ($D = 2$) and spherical ($D = 3$) domains. (A second reason that D is non-integer is that it is assumed in the theory that domain wall velocity is independent of domain radius, which is definitely not true.) The second parameter is the product of the maximum leakage current density i_m and time at which displacement current peaks t_{max}, divided by P_s to give a dimensionless ratio:

$$\frac{i_m t_{max}}{P_s} = \Lambda . \tag{6.10}$$

The constant Λ must satisfy certain self-consistent constraints within the Ishibashi theory. Typically $\Lambda = 2.0$ for three-dimensional domains ($D = 3$) and $\Lambda = 1.65$ for domain dimensions averaging 2.5. Data for PZT show that this model gives self-consistent values of D and Λ.

When experimental data are fitted to the formula in (5.14) it is necessary in general to subtract the non-switching current transients from the switching transients. This is seen most easily be referring to the hysteresis curve in Chap. 1 (Fig. 1.1). The change in displacement vector D (or polarization P) for a ferroelectric initially in a negative polarization state and subjected to a positive switching voltage is called P in testing jargon (for *positive switching pulse*). By comparison, the change in polarization for a ferroelectric initially in a positive polarization state and subject to a positive (non-switching) voltage is called U (for *up voltage response*). The U response is just the response of a linear dielectric, but if the dielectric constant ε is large, this can be a significant fraction of the P response. Similarly, negative voltages applied to switch the polarization of a ferroelectric initially polarized positive yield an N response (*negative switching pulse*); and applying a negative voltage to a film already negatively polarized results in a D (*down*) response. In general the P response differs from N, and U differs from D This can be due to horizontal shifts in the hysteresis curve along the field (or voltage) axis, due to built-in bias fields (e.g., due to charged defects), or along the vertical polarization axis, due to inequivalent top and bottom electrodes and resultant space-charge formation at the electrode–ferroelectric interfaces.

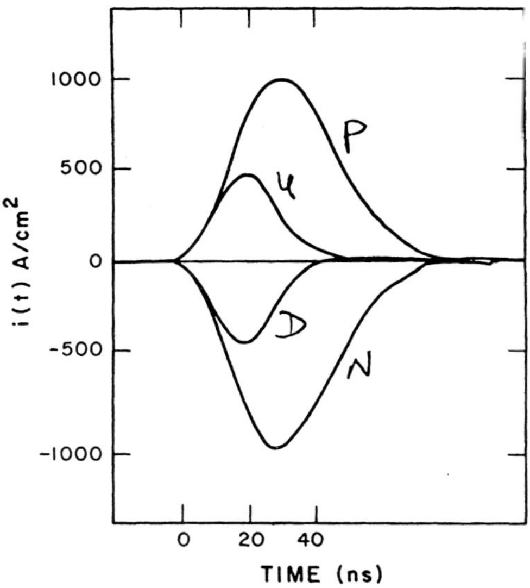

Fig. 6.9. Current transients P, U, N, D, as discussed in the text for PZT films

The P, U, N, D current transients $i(t)$ are shown for PZT in Fig. 6.9. Note that for the first 15 ns the U and P traces are identical. Thus, the sense amplifiers cannot distinguish between a positively polarized cell and a negatively polarized cell for this time (i.e., they cannot discriminate between a stored 1 and a stored 0 in the Boolean algebra of the random access memory). Figures 6.10 and 6.11 show the shift in the switching response $i(t)$ at different temperatures and at applied voltages in PZT films.

It is useful to relate the time $t_{\max}$ at which the displacement current peaks to both temperature and voltage (or field). Figure 6.10 shows that for PZT the temperature dependence is given by

$$t_{\max} = t_{\max}(0)\tau^m , \tag{6.11}$$

where $t_{\max}(0)$ is $t_{\max}$ at $T = 0$, τ is reduced temperature

$$\tau = \frac{T - T_c}{T_c} , \tag{6.12}$$

and m is an empirical number ($m = 1.7$ in PZT).

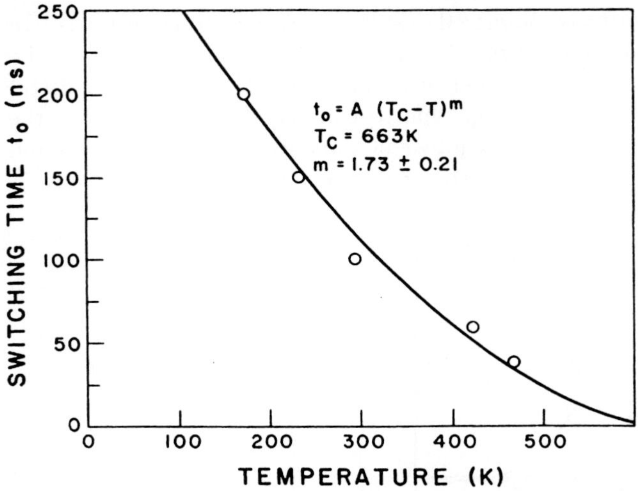

Fig. 6.10. Switching time $t_{\max}$ versus temperature in PZT films [265]

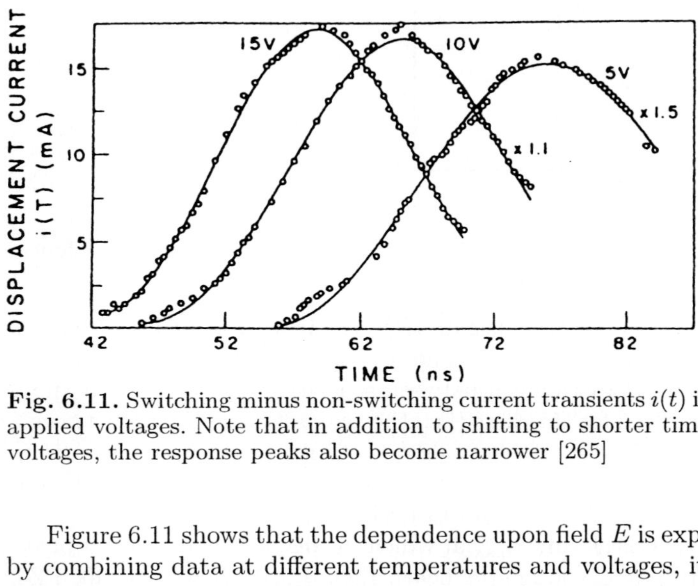

Fig. 6.11. Switching minus non-switching current transients $i(t)$ in PZT at different applied voltages. Note that in addition to shifting to shorter times with increasing voltages, the response peaks also become narrower [265]

Figure 6.11 shows that the dependence upon field E is exponential. In fact, by combining data at different temperatures and voltages, it was possible to determine that the switching time $t_{\max}$ can be expressed at all voltages and temperatures in terms of a single reduced parameter τ/V. This is shown in Fig. 6.12 for PZT. When expressed in this way, the switching time $t_{\max}$ has the form

$$t_{\max} = t_{\max}(0) \exp \frac{\alpha \tau}{E} , \tag{6.13}$$

where α is the activation field for switching at $T = 0$ (kelvin). For PZT α is typically 20 MV/m.

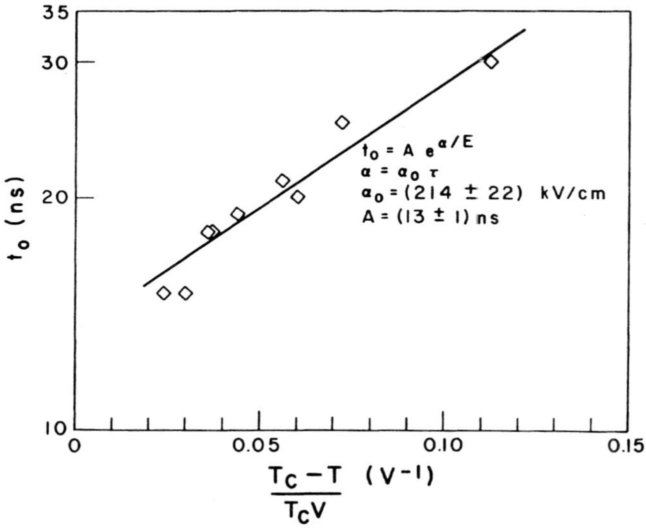

Fig. 6.12. Switching time $t_{\max}$ versus reduced temperature–voltage parameter τ/V, showing that data at all temperatures and fields scale

Equation (6.13) was found [266] to be more accurate than earlier expressions [267] for $t_{\max}$,

$$t_{\max} = \text{constant } E^{-j} \tau^k \,, \tag{6.14}$$

and in addition it explains why j and k were always found empirically to be about 1 or 2: these are simply the average of the leading linear and quadratic terms if the exponential in (6.13) is expanded for large E. It is also useful to note that the empirical relationship between spontaneous polarization and α,

$$\alpha = \text{constant } \frac{P_s^3}{T} \,, \tag{6.15}$$

can be derived starting with (6.13) and using the mean field expression for $P_s(T)$ in (1.3). It is interesting to note parenthetically that (6.15) predicts that it becomes harder to switch ferroelectrics as temperature is reduced to zero.

Most switching data on PZT or SBT films have used the Ishibashi theory for fitting parameters. However, it should be noted [269] that this theory, based upon the Avrami model of crystal growth [270], is not valid for finite grains. One interesting fact related to this is the observation [271] that switching time decreases with increasing capacitor area. This observation stimulated the modification of Ishibashi's model to include preexisting nuclei of domains of a finite initial size [263, 272, 273].

Duiker's finite-grain model [263] gives an explicit prediction that $i(t)$ is maximum at a time

$$t_{\max} = \frac{1}{v}\left[\left(\frac{vD}{Rk}\right)^{1/(D+1)} - r_{\mathrm{c}}\right] , \tag{6.16}$$

where k is a geometric factor equal to $4\pi/3$ for $D = 3$. For PZT $D = 3$ is a good approximation [265] (compared with $D = 1$ in uniaxial SBT). If we neglect the critical radius r_{c} in (6.16) and use values for R and v known from fitting current transient shapes $i(t)$, we predict

$$t_{\max} = 0.6 \pm 0.1 \text{ ns} , \tag{6.17}$$

which is in good agreement with the fastest switching time ever reported for PZT of 0.38 ns.

Note that coercive fields (and probably switching times) are lower for domains in the center of grains than near the grain boundaries [274].

In closing this chapter on switching kinetics, it is useful to point out what one should not do in modelling domain switching. Some authors have postulated that domain wall velocities v are linearly proportional to applied field E, assuming from Lines and Glass a kind of effective viscosity and viscous-drag-limited kinetics. Then they set $E = -\nabla\Phi$, where Φ is the solution to Laplace's equation, so that $v = -\beta\nabla\Phi$. In general, both steps are wrong. Domain wall velocities are proportional to $E^{3/2}$ in some field regimes (Stadler) and $\exp(-\alpha/E)$ in others. They are rarely linear in E. Furthermore, Laplace's equation is invalid in situations that have rapidly changing fields, such as a ferroelectric thin film subjected to a fast rise-time (ns) voltage pulse.

7. Charge Injection and Fatigue

Fatigue is the characteristic of ferroelectrics that the amount of charge q that is switched with repetitive bipolar applied voltages decreases with cycling. This is usually described as a decrease in remanent polarization $P_r(N)$ with switching cycles N. However, in reality the polarization per primitive crystallographic unit cell remains unchanged with N. Instead, the number of cells and domains that switch decreases slowly with increasing N, as domain walls become pinned. This pinning can occur at grain boundaries, at surfaces, or at other defect sites, particularly extended defects.

It is possible to confirm these ideas by *depinning* the trapped domain walls [275] as shown in Fig. 7.1. This recovers 100% of the original switched charge and hence recovers the original virgin squareness of the hysteresis loops.

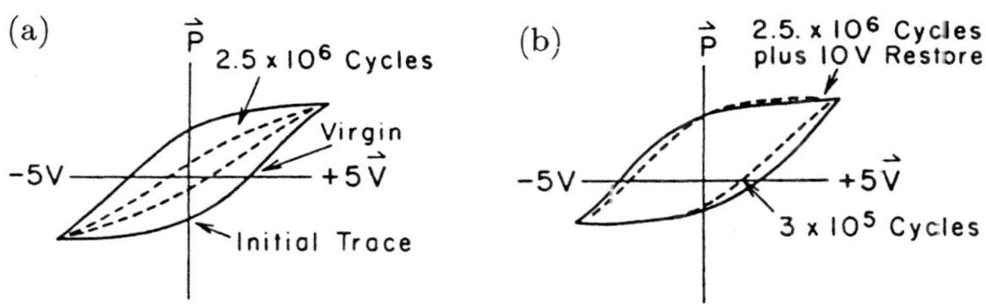

Fig. 7.1. (**a**) Hysteresis curves in KNO_3 before and after fatigue (bipolar switching cycles). (**b**) After a single 10 V depinning 'restore' pulse the hysteresis returns to its virgin (unswitched) value [275]

Fatigue is a complicated electrochemical process. The point of view in this book is that it arises in part from charge injection at the electrode–ferroelectric interface (see Chap. 10 and the discussion concerning Fig. 10.7) and is therefore not an intrinsic property of the bulk ferroelectric material per se. (It is useful to keep in mind that for a metal electrode with a 4 to 5 eV work function, application of a 10–100 MV/m electric field results in an injected current of several $\mu A/cm^2$.) The lack of fatigue in SBT compared with PZT is thought to arise from the fact that domain walls have much smaller pinning energies in SBT, exhibit rather shallow traps, and have a lower concentration of oxygen vacancies near the electrode interface [276,

277, 278]. The lower oxygen vacancy concentration is thought to arise from the fact that the molecular layers adjacent to the electrode are Bi_2O_2 and not TaO_6 octahedra [279]. Rather curiously, small additions of bismuth to PZT also greatly reduces fatigue, according to Lee and Lee [280].

Recent studies of fatigue [280] have shown very clearly that although fatigue results from pinning of domains, this pinning does not occur randomly throughout the interior of the film; rather, it occurs under normal switching conditions at the interface only. By cycling at extremely low frequencies (1.7×10^{-3} Hz) it is possible to separate out the 'bulk' pinning in the interior of the film from pinning at the electrode interfaces.

Detailed models of domain wall pinning in PZT films differ in their assumptions of whether it is possible for domain walls to cross grain boundaries. The work of Colla et al. [281] assumes that this is impossible, whereas Gruverman et al. [282] have shown directly via Atomic Force Microscopy (AFM) that it readily occurs. Colla et al. have discovered a rather interesting non-monotonic dependence of switched charge upon bipolar cycles (fatigue). This self-recovery depends upon the magnitude of applied fields in the range $2E_c < E < 4E_c$ and is interpreted as inhibition of residual *seeds* (domains antiparallel to the majority). This effect seems closely related to imprint.

Fatigue also results directly in resistance degradation (and eventually, to time dependent dielectric breakdown). It was first shown [283] by Melnick et al. (Fig. 4.4) that the resistivity of fatigued PZT drops significantly after bipolar voltage stressing. In addition they showed that this effect was due to the surface/interface term B in (4.7) and not the bulk term A_0. For BST it was very recently established [284] by Zafar et al. that unipolar voltage stressing in BST/Pt does the same thing; that is, prolonged application of a dc voltage decreases resistivity by reducing the Schottky barrier height (and/or width). Zafar et al. find that the barrier height reduces a critical amount (0.18 eV) with dc voltage duration. It may be significant that this is approximately the amount by which a monolayer of deposited oxygen on the Pt surface will change the Pt electrode work function [285]. Thus, stabilizing the Schottky barrier height with time is one of the most important engineering chores for both ferroelectric DRAMs and NV-RAMs. The use of oxide electrodes such as IrO_2 or YBCO probably does this electrochemically.

It is my view that both fatigue and TDDB are controlled primarily by electrode–ferroelectric interface effects and not the bulk properties of the dielectric, although the transport through the bulk is of obvious importance [286].

7.1 Model of Dawber

Although concensus of opinion at present seems to be that fatigue in ferroelectric thin films is primarily due to charge injection and that Arlt's 1988 model of preferential site diffusion for oxygen vacancies [287] is applicable only to bulk specimens, Dawber and Scott have very recently published [288] a model of thin-film fatigue that is free from the requirement of electron injection from the cathode. The basic model uses Arlt's idea plus Brennan's ansatz [289] that fatigue arises from the number of defects in planes perpendicular to the polarization[1]. In bulk the dominant effect is the distribution of oxygen vacancies between different lattice sites; in thin films the dominant effect is due to build up of charged vacancies at the metal–ferroelectric interface.

Arlt's model consists of a TiO_6 oxygen octahedron in a perovskite lattice with one oxygen vacancy, randomly occupying site numbers 1 or 3, at the $+P$ or $-P$ cube faces ($+z$ or $-z$ polar axis), or sites 2, 4, 5, 6 in the plane of the nonpolar axes. These site occupancies are modeled with a triple-potential well. The key parameters in the model are the activation energy W_0 and the energy level splitting ΔW caused by the spontaneous polarization and accompanying strain.

The oxygen vacancy and the opposing oxygen ion form a reorientable dipole that can be aligned in an external field via ion vacancy hopping from site 1 to site 3. The rate equation for this system is

$$\frac{\mathrm{d}\boldsymbol{n}}{\mathrm{d}t} = -A\boldsymbol{n} \,, \tag{7.1}$$

where A is a 6×6 matrix with elements

$$a_{ij} = a_0 \exp\left(-\frac{W_{ij}}{k_\mathrm{B}T}\right) . \tag{7.2}$$

Rotational symmetry requires $n_2 = n_4 = n_5 = n_6$. The built-in electric field resulting from these defect dipoles is given by

$$E_\mathrm{bi} = \frac{\Delta n(t)\Delta W}{P_\mathrm{s}} \,, \tag{7.3}$$

and will be opposite to the field that originally poled the ferroelectric when $\Delta n(t) = n_3(t) - n_1(t)$. The solution to these equations is

$$\Delta n(t) = D_1 \exp(-\nu_1 t) + D_2 \exp(-\nu_2 t) + n_3(\infty) - n_1(\infty) . \tag{7.4}$$

Both D_1 and D_2 depend upon the initial oxygen vacancy concentration and the saturation vacancy densities.

Arlt has also applied this model to explain de-aging (ac signal) to remove the original internal bias field. Over a complete hysteresis cycle the coefficients average to give

$$\langle a_{12}\rangle = \langle a_{32}\rangle = \nu_0 \cosh \Delta W/kT . \tag{7.5}$$

---

[1] This predicts a strong anisotropy in fatigue, depending upon film orientation. This has been confirmed by fatigue measurements on (111) and (001) PZT by Clive Randall et al. at Penn State University (in press), which show that the ratio of fatigue cycles required $N(111)/N(001) = 10^5$.

This term is used by Dawber and Scott (Integ. Ferroelec., in press) to generate expressions for $P(N)$ fatigue in bulk ferroelectrics. However, for thin films the fatigue process is modelled by the accumulation of oxygen vacancies in the region near the electrode. Self-ordering of oxygen vacancies at high concentrations is a well-documented phenomenon in perovskites see Sect. 7.2 and is used by Dawber and Scott to obtain an analytic expression for fatigue. In ferroelectric thin films the oxygen vacancies are produced in deposition and the concentration varies a great deal throughout the film [290, 291]. Oxygen vacancies are highly mobile under an applied electric field, so that the oxygen vacancy distribution changes throughout the fatigue process. As oxygen vacancies build up at the interface they can order themselves into domain-wall-pinning defect planes. Mihara et al. [292] found an oxygen vacancy concentration near the surface of 5×10^{20} cm^{-3} and 10^{18} cm^{-3} in the interior of the film. Dey [291] has measured a space charge distribution which gives a concentration of 5×10^{20} cm^{-3} at ca. 10 nm from the interface. The interfacial concentrations are sufficient that one might expect to see self-ordering of oxygen vacancies in the near-interfacial regions. The following set of equations calculates the accumulation of oxygen vacancies in a layer near to the metal–ferroelectric interface, where the increased concentrations could lead to a self-ordering of vacancies that may pin domain walls. The following derivation is based on a system in which there is a single depletion region of about 20 nm at the reverse-biased junction. There is no significant depletion region at the forward-biased junction, and it is assumed that all the voltage falls linearly across the depletion region; hence $E = V/(20\ \text{nm})$. Scott [293], Waser [286], and Dey [291] give further discussion on this point. An oxygen vacancy, being analogous to a spherical cavity, will feel a modified electric field

$$E_{\text{cavity}} = E\left(\frac{3\varepsilon_{\text{s}}}{2\varepsilon_{\text{s}}+1}\right) \tag{7.6}$$

and when ε_{s} is large, as it is here (approximately $400\varepsilon_0$), $E_{\text{cavity}} = 3E/2$.

The flux density under forward dc field is given by

$$j = \mu E_{\text{cavity}} D \exp\left(\frac{\Delta S}{k}\right) \exp\left(\frac{zqbE_{\text{cavity}}}{2kT}\right) \text{m}^{-2}\text{s}^{-1}\,. \tag{7.7}$$

Here m = defect migration mobility [294]; D = defect concentration; ΔS = entropy for defect movement; z = valence of defect; q = electronic charge; b = jump distance for defect (interatomic spacing).

As a sample is fatigued, the bias is alternated between forward and reverse. The flux of oxygen vacancies would be towards the junction under reverse bias and away from it under forward bias, except for the fact that the electric field is much smaller under forward bias than reverse bias, so that we may neglect the forward bias oxygen vacancy flux. The flux of oxygen vacancies then becomes

$$j = \mu \frac{3}{2} E D \exp\left(\frac{\Delta S}{k}\right) \exp\left(\frac{3zqbE}{4kT}\right) \mathrm{m}^{-2}\mathrm{s}^{-1} . \tag{7.8}$$

Neglecting duty-cycle effects, there will always be one junction under reverse bias, and so there is always a flow of oxygen vacancies to one or the other of the junctions at any time that an AC voltage is applied. We convert this flux to the rate of change of the concentration of oxygen vacancies (D) [295] multiplying by cross-section over volume.

$$\frac{\mathrm{d}D}{\mathrm{d}t} = \frac{3\mu E D}{2L} \exp\left(\frac{\Delta S}{k}\right) \exp\left(\frac{3zqbE}{4kT}\right) \mathrm{m}^{-3}\mathrm{s}^{-1} , \tag{7.9}$$

where $a = zqb/2kT$ and L is the thickness of the film. Integrating with respect to time and substituting $t = N/f$ (i.e., neglecting duty-cycle effects), we obtain:

$$\frac{D(N)}{D(0)} = \exp\left[\frac{3}{2}\frac{\mu E}{L} \exp\left(\frac{\Delta S}{k}\right) \exp\left(\frac{3a}{2}E\right) \frac{N}{f}\right] . \tag{7.10}$$

Assuming that the switchable polarisation is inversely proportional to the concentration of vacancies near the interface,

$$P(N) = A \exp\left[-\frac{3}{2}\frac{\mu E}{L} \exp\left(\frac{\Delta S}{k}\right) \exp\left(\frac{3a}{2}E\right) \frac{N}{f}\right] + B . \tag{7.11}$$

In the above equation A and B represent the fraction of charge that can be pinned by the fatigue mechanism and are the only adjustable parameters in our model. In our calculations we have used the mobility measured by Maier, which at 500 K was $10^{-9}\mathrm{cm}^2/\mathrm{Vs}$. The temperature dependence of this mobility is of the form $\exp(-c/kT)$ where c is the activation energy for oxygen vacancy movement and is around 0.9–1.1 eV. The jump distance is equivalent to the distance between two nearest neighbour oxygen vacancies which is about 0.3 nm. The relevant change in entropy is that associated with the self-ordering phase transition described in Sect. 7.2 . It may be described by an activation energy expression

$$\exp\left(\frac{\Delta S}{k}\right) \propto \exp\left(\frac{H_\mathrm{a}}{kT}\right) . \tag{7.12}$$

That H_a is the relevant activation energy may be seen from the manner in which it goes to zero when the phase transition occurs. Estimating the unit cell as being 5Å across we find that the density of oxygen ions (three per unit cell) is $2.40 \times 10^{22}\mathrm{cm}^{-3}$. If the relevant oxygen vacancy concentration is $5 \times 10^{20}\mathrm{cm}^{-3}$, then the oxygen vacancy concentration as a percentage of oxygen sites is 2.10%. Steinsvik finds that H_a for this situation is 0.7 eV.

The numerical predictions of (7.11) are compared in Fig 7.2 with the experimental PZT data at different fields of Mihara et al. Agreement is excellent. In Fig. 7.3 we compare the dependence of switched charge versus fatigue cycles at 5 V and 323 K with the data [296] of Colla et al. (1998) at

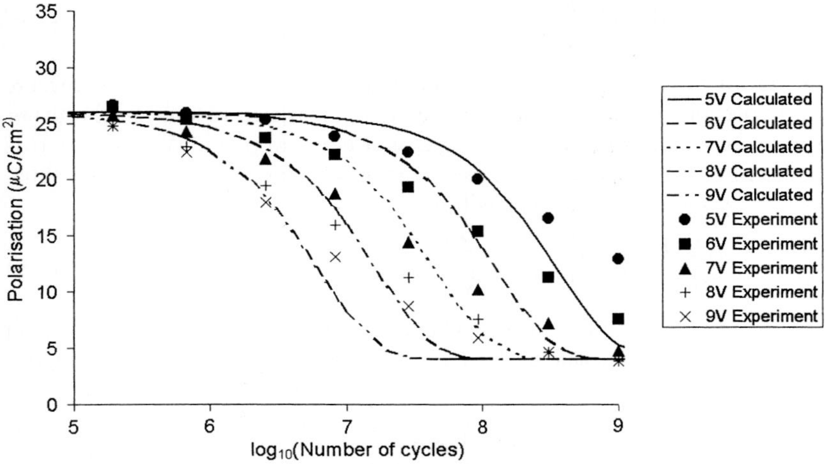

Fig. 7.2. PZT fatigue at different applied voltages: points represent experiment (Mihara et al.); lines represent the theoretical prediction (7.11)

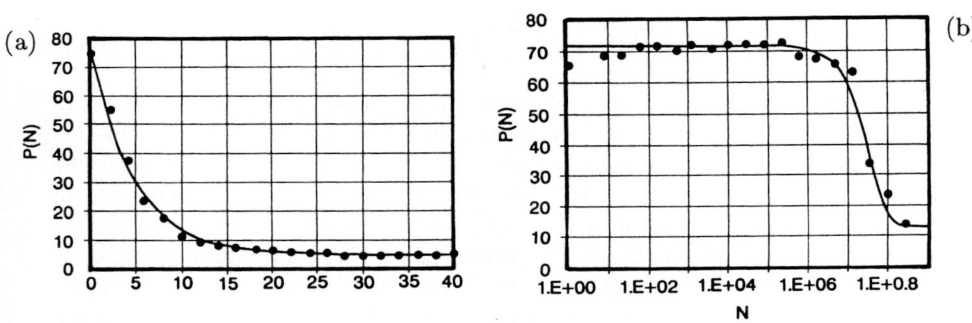

Fig. 7.3. (**a**) PZT fatigue at low frequencies (17 mHz) and (**b**) at high frequencies (30 kHz), from Colla et al. [296]. Solid curves are theoretical calculations from (7.11) with no adjustable parameters except B

17 mHz (sub-Hz data) and at 30 kHz. Again the agreement is excellent. Note that the mHz and kHz data are both correctly predicted by our model.

Fatigue over a wider range of frequencies, with experimental data [297] from Moritomo et al. [297], is quantitatively explained by the model (7.11) with no adjustable parameters. The temperature dependence is more subtle, due to T-dependence of several competing terms in (7.11) and is in excellent agreement with the data of Paton et al. in Fig. 7.4 [297]. The main term is the ionic mobility μ, which Zafar et al. showed to increase by several orders of magnitude as T increases [294].

In conclusion, this model shows that it is not necessary to include charge injection to model fatigue in PZT. Based upon earlier work by Arlt and by Brennan, Dawber has given an analytic expression for $P(N, E, f)$ in (7.11)

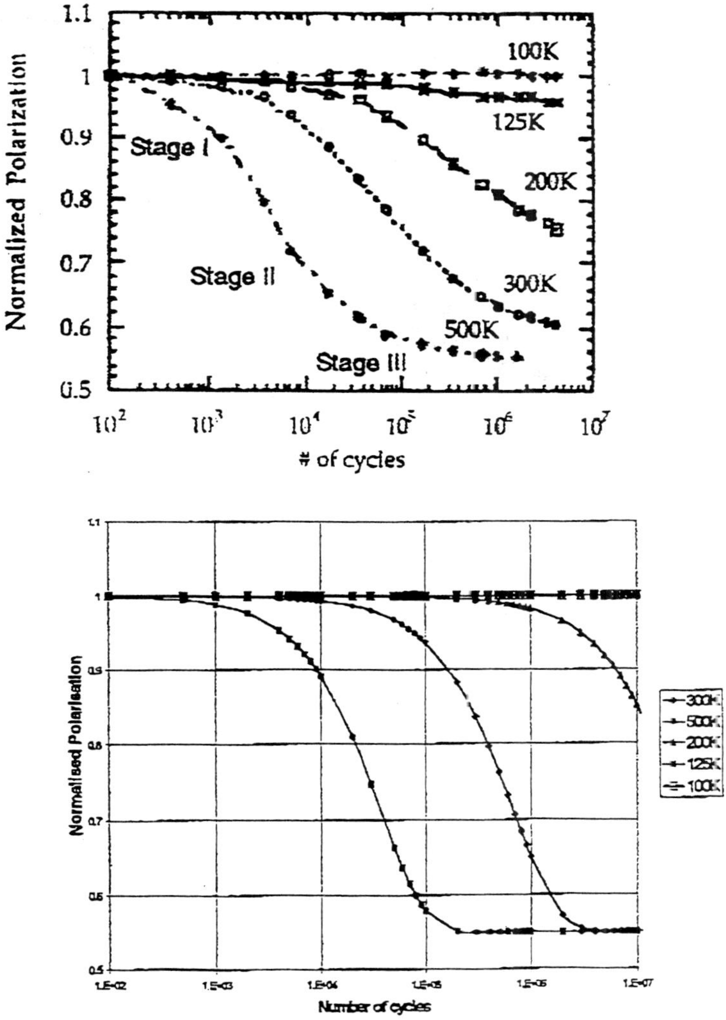

Fig. 7.4. Temperature dependence of fatigue in PZT

that works well as a device equation. This supports earlier refutation [298] by Tagantsev et al. (1995) of the conclusions [299] of Larsen (1994).

The basic results of Colla et al. show that domain walls become pinned and still exist under large applied voltages. This is observed only at extremely low frequencies and results in an increase in dielectric constant at such sub-Hz frequencies, due to the domain wall contribution initially shown in the classic work of Fouskova [300]. Colla et al. concluded that fatigue is dominated by inhibition of domain nucleation at the electrode surface. This is also supported by the work [301] of Majunder et al. (1996), who showed that their hysteresis 'tilting' arises from the increase in width of a non-switching layer at the electrode–ferroelectric interface. Very recently Taylor et al. [301] modeled the slow (sub-Hz) switching of domains in PZT and concluded that slow switching pins domain walls throughout the interior of the film via motion of charged mobile defects, whereas rapid (kHz) switching results in an inhibition of nucleation only at the elect rode–ferroelectric interface.

Primary support for charge-injection models of fatigue comes from I-Wei Chen, who showed [302] that fatigue in PZT films is greatly reduced with p-Si electrodes, compared with n-Si electrodes, due to the elimination of electron injection from the cathodic electrode.

Finally, the role of grain size on fatigue, as made clear in the Sc-doping experiments on PZT by Yoo and Desu, must be examined further.

At this stage of analysis, the oxygen vacancy models of Arlt–Brennan–Dawber, the charge injection contribution emphasized by I-Wei Chen, and the grain-size contribution illustrated by Yoo and Desu all appear to be important facets of fatigue. No comprehensive theory including all these effects has yet been presented to yield device models of $P(N, E, f)$.

7.2 Oxygen Vacancy Ordering as a Fatigue Mechanism in Perovskite Ferroelectrics

It has been known for twenty years [303, 304, 305, 306, 307] that increasing numbers of oxygen vacancies in perovskites cause a self-ordering into planar structures. The context of these observations, however, was that of the study of bulk materials in which the vacancy concentration was controlled via doping (e.g., with Fe substitutional for Ti) or quench temperature. More recently and quite independently, Arlt [308] and Brennan [309] have modeled fatigue in oxide perovskites with a similar concept involving preferential electromigration of oxygen vacancies into sites in planes parallel to the electrodes, and both Yoo and Desu [310] and we [311] have shown that such models can account quantitatively for the known but previously unexplained dependence of fatigue [switched charge $P(N)$ versus switching event number N] upon frequency [312], temperature [313], and applied electric field [314]. Until now, however, there has been no direct microscopic confirmation of the ordered planes of oxygen vacancies we postulate in our model. A main aim of the present section is to propose detailed experiments using NMR, EELS, and Mossbauer effect studies to test this proposed mechanism. To our knowledge this is the first proposed use of Mossbauer techniques to analyze fatigue in ferroelectrics.

Previous studies of oxygen vacancy ordering in perovskites include a $Ba_2In_2O_5$ study by Zhang and Smyth [315] and studies of calcium titanate/ferrites by Grenier et al. [303, 304, 305] and most recently, Becerro et al. [306]. In these systems it is observed that in dilute amounts oxygen vacancies are uncorrelated point defects, but that with increasing vacancy concentrations (or lowered temperature T) the vacancies cluster, initially in one-dimensional chains whose lengths increase with vacancy concentration (or decreasing T), and above a certain vacancy density, in planes of specific geometries.

Figure 7.5 illustrates [306] the gradual growth from an isolated point defect to a linear chain of oxygen vacancies along [101] in the cubic perovskite structure.

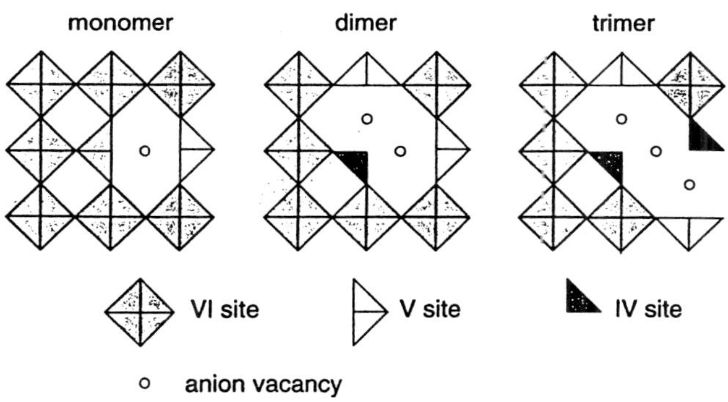

Fig. 7.5. Projection of the cubic perovskite structure onto the $(0k0)$ plane, showing the growth of an oxygen vacancy chain along (101) from an isolated vacancy [306]. The Fe valence sites +6, +5 and +4 are marked

The connection in the present paper is to link these established observations with the hypotheses presented by Arlt et al. and by Brennan that a similar [2D] planar self-organization of oxygen vacancies develops with applied voltage cycling in ferroelectric perovskite thin-film memory capacitors. It is specifically proposed that ^{57}Fe Mossbauer studies or yttrium-NMR can be employed to verify the structures suggested. As shown by Beccero et al., the signature of the [2D] planar arrays of ordered oxygen vacancies in titanate perovskites is the abrupt disappearance of any Fe^{+5}-valence Mossbauer signal (since those arise only from the ends of finite chains of oxygen vacancies, which become a negligible percentage of all Fe as the chains lengthen and form planar arrays), concomitant with the occurrence of bright planes in the high resolution TEM patterns (HREM).

It is useful to consider driving mechanisms for this process in real capacitors. These will include metal atoms diffused into the ferroelectric from the electrode, as well as charge injection of electrons from the cathode with consequent increase in oxygen vacancy concentration (impact ionization and related processes). In our recent quantitative model of fatigue [311], the high-field transition from Schottky-dominated thermionic emission to Fowler–Nordheim tunneling [316] at around 350 kV/cm plays no role in itself in changing the rate of fatigue, contrary to the suggestion of Stolichnov and

Tagantsev [317], except in as much as it results in an increase in leakage current J. That is, the mechanism (Fowler–Nordheim) per se is of no importance whatsoever. This is supported by rather detailed studies by Nordby's group [318, 319, 320].

We point out explicitly that these ideas lead to modeling the onset of fatigue in perovskite oxides as closely analogous to a structural phase transition (viz. perovskite to brownmillerite). This analogy will be developed in more detail elsewhere, complete with order parameter, conjugate field, critical (universal) exponents, etc. In this approach it is not the oxygen vacancy concentration per se that is the order parameter, but rather the ordered plane of vacancies. Thus the correlation length λ and its characteristic exponent ν are important. The dependence of fatigue upon temperature was first shown for lead zirconate-titanate (PZT) by Eaton at Ramtron [321] and extended to higher temperatures by Mansour's group at Purdue [313]. The demonstration that fatigue affected both optical phonons in the 200–300 cm^{-1} region and also polaron absorption associated with oxygen vacancies in the 4 000 cm^{-1} region was first made by Guettler et al. [322].

We note parenthetically that the oxygen vacancy ordering mechanism we propose here for perovskite titanates such as PZT is inoperative in Aurivillius layer structures such as $SrBi_2Ta_2O_9$ (SBT) since the oxygen vacancies in that system prefer sites in the Bi_2O_2 planes, where their effect upon the polarization is negligible, and not within the $SrTaO_3$ octahedra that control ferroelectric switched polarization from Ta–O bonds. That is a primary reason why SBT does not fatigue [323].

If the oxygen vacancy ordering proposed here as a fatigue mechanism always required 50% vacancies (as in the perovskite–brownmillerite phase transition), then it would only be of academic interest and not really pertinent to fatigue in PZT. However, Steinsvik et al. [318] find that something more important happens at 7% vacancy concentration – an ordered $2 \times 2 \times 1$ superlattice. They find that the activation enthalpy H for annihilation of oxygen vacancies is 2.0 eV for stoichiometric ABO_3; 1.0 eV for $ABO_{2.95}$; 0.5 eV for $ABO_{2.90}$; zero for $ABO_{2.80}$; and −0.2 eV for $ABO_{2.60}$, thus implying a phase transition at about 7% oxygen vacancy concentration ($ABO_{2.80}$). Equally important, they point out that empty electron states on some oxygen atoms are signalled in EELS spectra by a shoulder at 527 eV on the oxygen K-edge, a microscopic test of this hypothesis.

This particular $2 \times 2 \times 1$ superlattice ordering of vacancies has not been modelled theoretically. However, Khachaturyan [307] has provided a detailed model for analogous centered cubic structures in which sublattice ordering occurs among octahedral interstices which become filled (e.g., with oxygen ions). In this model there is a self-ordering of magnitude $1 - \alpha$ for oxygens along x and y and $1 + 2\alpha$ for oxygens along z, exactly as in the electromigration pattern of fatigued PZT assumed by Arlt. Khachaturyan's model is the opposite of that encountered in PZT fatigue in that he starts with a body-centered cubic lattice of metal ions (which we take as Pb and Zr or Pb and Ti in PZT) and mathematically adds oxygen ions at octahedral sites around the Zr or Ti to produce ordered superlattices. By comparison, in PZT fatigue we start with octahedral sites initially all filled with oxygen ions and then substract them to form vacancy superlattices.

In Khachaturyan's model large populations ($\alpha = 1/3$) of oxygen ions in those interstices lead to the ordered superlattices known [324, 325] in Ta_2O and Nb_2O, analogous to Beccero's perovskite–brownmillerite transition at $\alpha = 1/2$.

8. Frequency Dependence

It is easiest to measure hysteresis loops like that in Fig. 1.1 at low frequencies. For convenience, line frequencies of 60 Hz (USA) or 50 Hz (Europe) are therefore most often employed. Alternatively, hysteresis is measured at 1 kHz. This gives consistent, reproducible values of coercive field E_c and remanent P_r and saturation polarization. However, these values will actually be frequency dependent (Fig. 8.1). Since real ferroelectric memory devices will have to operate at 100–500 MHz in the near future, the differences between E_c (60 Hz) and E_c (300 MHz) may be very large and very important.

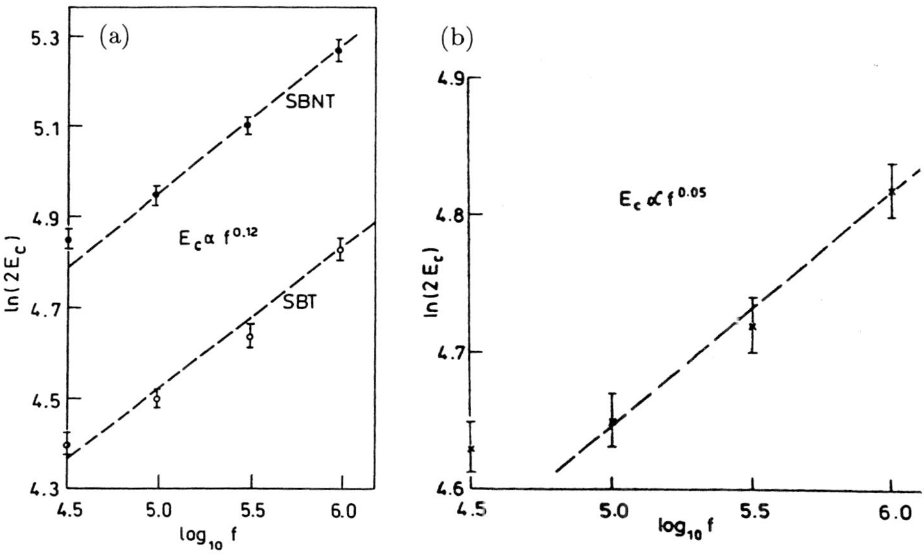

Fig. 8.1. Frequency dependence of coercive field in (**a**) SBT and SBTN at high frequencies and (**b**) PZT; mist deposition samples from Symetrix Corp [328]

The physical reason why coercive field varies with frequency f is that domain switching in a ferroelectric involves domain wall motion (except at extremely high fields or extremely small capacitor cells). Motion of domain walls through a ferroelectric involves a certain resistance or viscosity, and viscosities generally increase with increasing frequency. Therefore coercive fields will increase with frequency in ferroelectrics. As discussed earlier in this book (Chap. 2), the dependence of coercive field upon frequency is equivalent to the dependence upon pulse width. One is simply the Fourier transform of the other. Higher frequencies or shorter pulse widths result in higher E_c.

8.1 Ishibashi–Orihara Theory

The theory of Ishibashi and Orihara for domain growth [326] is based upon the analogous theory of Avrami [327] for crystallite growth. In this model the fractional area of the cell that is switched with time is an explicit function $A(E, f)$ of field E applied and frequency f:

$$A(E, f) = 1 - \exp\left[-f^D Y(E)\right] , \tag{8.1}$$

where D is the domain dimension and Y is a function containing the field dependence. Under certain approximations, $Y(E)$ is a power law:

$$Y(E) = \text{constant } E^k , \tag{8.2}$$

with k approximately equal to 6.

This theory was first tested for SBT films [328] using data (Figs. 8.2 and 8.3) from DeVilbis et al. [329] for frequencies up to 2 MHz. Assuming $D = 1$ for highly uniaxial SBT, this gave a theoretical prediction of

$$E_c(f) = f^n = f^{1/6} , \tag{8.3}$$

which compares very well with the experimental value $n = D/k = 0.12 \pm 0.02$.

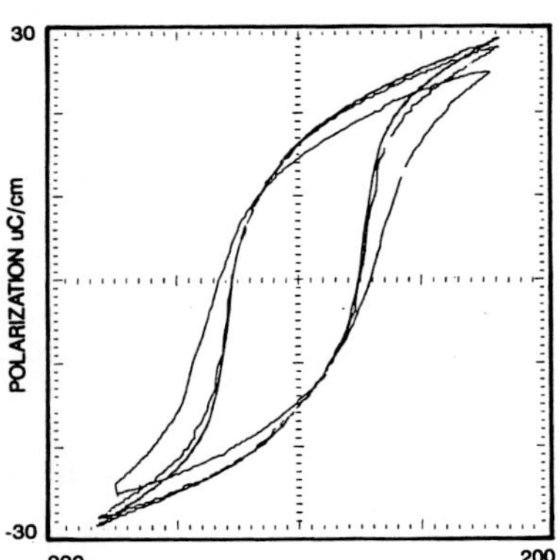

Fig. 8.2. Original unpublished $E_c(f)$ data on SBT from DeVilbis. From left to right the curves are at $\log_{10} f = 4.5$, 5.0, 5.5, and 6.0 (1 MHz), yielding $n = 0.12$

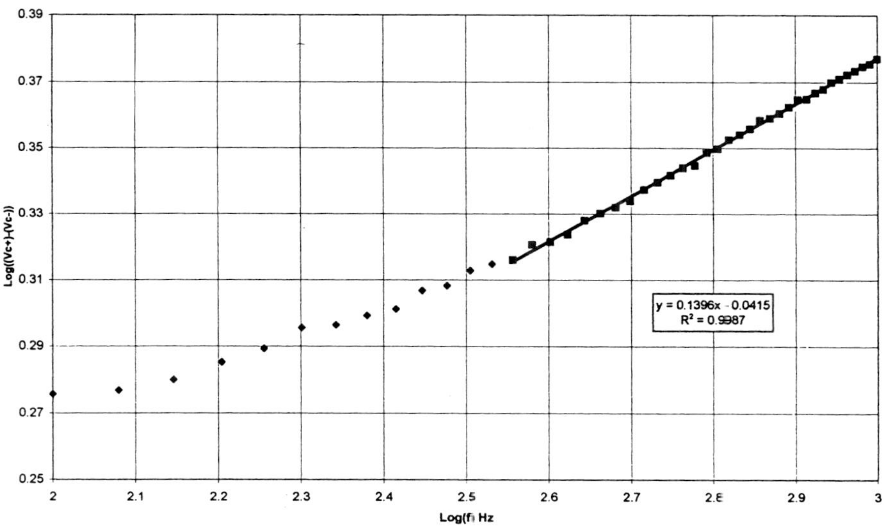

Fig. 8.3. Data from 10 Hz to 1 kHz (Dawber and Scott), showing the same exponent of 0.13 above 150 Hz but flat, nearly quadratic behaviour below

Subsequently this result was confirmed independently by Ramesh [330] and extended to very low frequencies (Hz) by Waser's group [331] and by Dawber and Scott [331]. The latter obtained a value of $D/k = 0.13 \pm 0.01$ from 10 Hz to 1 kHz, thus extending DeVilbis' work to much lower frequencies with the same exponent.

Recently Scott and Dawber have demonstrated a rather rapid charge injection process in SBT with only a few switching cycles at low frequencies (around 30 Hz). The bias field grows 10–20% from an initial low value as the capacitor is cycled.

8.2 Interface Effects

Zafar et al. have shown [332] that the capacitance in BST/Pt thin films has two parts (as we showed earlier [333]; see the discussion in Chap. 13, Fig. 13.8). The interfacial capacitance is independent of frequency, and the bulk dielectric constant has a power law dependence upon frequency. The exponent of the power law is approximately $f^{1/m}$ with $m = 120$. There is no known theoretical relationship between this frequency exponent and the frequency exponent for the coercive field.

9. Phase Sequences in Processing

PZT and SBT exist in several different phases. When PZT films are deposited under normal conditions the phase that is formed has the pyrochlore structure. These pyrochlores are generally ferroelectric but only at cryogenic temperatures; thus, they are not useful for room temperature ferroelectric memory devices. The pyrochlores deposited by sputtering, sol–gel spin-on, or other techniques are subsequently subjected to an annealing cycle, either in a furnace (typically an hour at 650°C) or via RTA (Rapid Thermal Annealing), typically for 90 seconds to a maximum temperature of about 800°C. These heat treatments result in 100% of the material being converted to the ferroelectric perovskite phase. Usually this heating process is carried out in an oxygen atmosphere to prevent oxygen loss from the film. The lowest temperature at which 100% of the film has been successfully converted to perovskite is 450°C [334]. Recently it has been suggested [334] that the pyrochlore phase is stabilized by Pb^{4+} ions in PZT, and efforts are underway to reduce this Pb^{4+} concentration to see if even lower temperature conversion to the perovskite phase is possible.

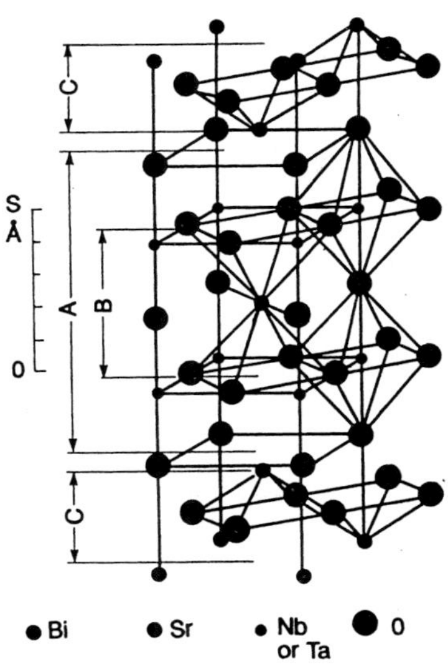

Fig. 9.1. $SrBi_2Ta_2O_9$ structure [338, 339]

SBT has a different sequence of phases. As-deposited films have a defective fluorite structure [335]. An ideal fluorite structure would have eightfold coordination of the Ta ion, with eight nearest-neighbour oxygens. The actual structure has octahedral sixfold coordination with two of these eight oxygen ions 'missing'. This fluorite phase is not ferroelectric. It must be heated to convert it to the ferroelectric Aurivillius layer-structure phase shown in Fig. 9.1. As with PZT, the heating can be done by furnace anneal or by RTA. Furnace anneals require a minimum temperature of about 600°C. (Lower values of temperature have been quoted for Plasma-Enhanced Chemical Vapour Deposition – PECVD – or CVD using cyclotron resonance, but the actual surface temperature is difficult to determine in these deposition schemes.) The fact that the processing temperature for SBT is thus about 150°C higher than for PZT is a significant disadvantage in commercializing SBT memories. The main difficulty is that it precludes the use of aluminium metallization for the SBT memory devices.

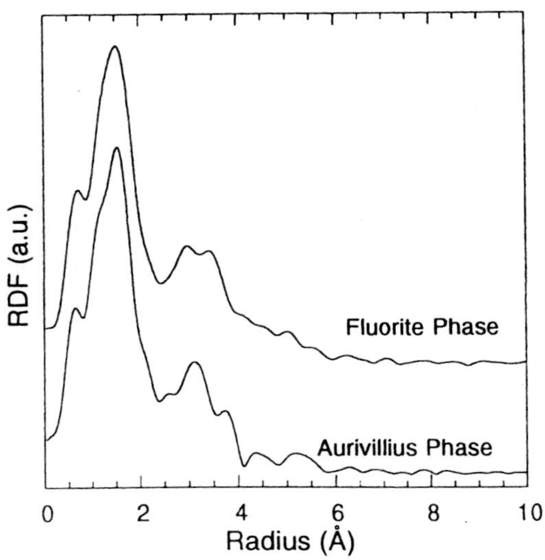

Fig. 9.2. EXAFS data for fluorite and Aurivillius phases of SBT [337]

It was proposed [336] that in the intermediate fluorite phase of SBT the Sr, Bi, and Ta ions randomly occupy the M-site of the basic MO_2 fluorite structure, despite the fact that their ionic radii and valences are different. This was confirmed [337] by the EXAFS data shown in Fig. 9.2.

The actual structure of SBT or SBN has a slight orthorhombic distortion from the tetragonal structure pictured [338, 339].

9.1 Sr-Deficient Optimized Films

Atsuki et al. [340] and Noguchi et al. [341] found that Sr-deficient SBT films (and SBTN films) have enhanced remanent polarizations, compared to strontium-stoichiometric SBT (or SBTN), but this remained unexplained until Watanabe et al. [342] showed in pure SBN ($SrBi_2Nb_2O_9$) that the Sr deficiency further randomizes the grain orientation, thus increasing P_r above its normal c-axis oriented value. As shown in Fig. 9.3, this is a 500% effect in SBN, compared to a 30% effect in SBT.

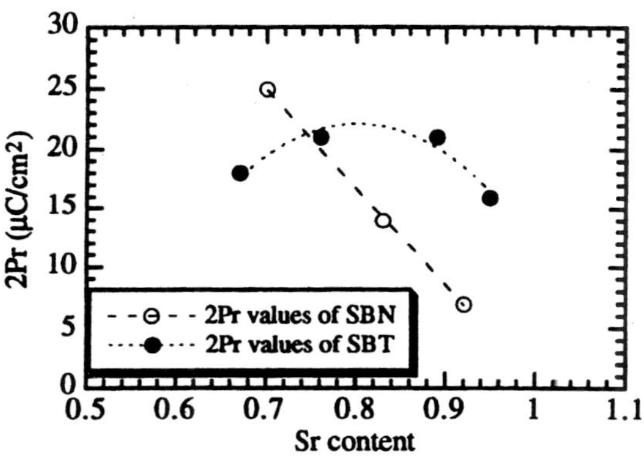

Fig. 9.3. P_r versus Sr concentration in SBT and SBN [342]

Lead diffuses easily in PZT and bismuth diffuses easily through SBT and reacts rather violently with platinum. This has given rise to a number of studies of mutual diffusion between ferroelectric thin films of these materials with their metallic electrodes. In general it is necessary to put a bismuth diffusion barrier between SBT films and the Si substrate.

9.2 The Role of Bismuth

Bismuth-containing ferroelectric films for memory applications have several problems or idiosyncrasies associated with the properties of bismuth. Firstly, bismuth is rather volatile and reacts violently with platinum. The electrical properties of Bi were first characterized accurately by Kapitza [343] and by Bussem et al. [344]. Later in the 1940s and 50s, they were studied in detail [345, 346, 347, 348, 349]. A good review is given by Colombani and Huet [350] who measured films down to 2.5 nm thickness. Bi is rare among metals in that it becomes more dense (like water) upon melting, a result thought to be due to covalent bonding [351]. Because of this covalent bonding, bismuth is almost always deposited as Bi_2 molecules; at room temperature the percentage of such molecules is nearly 100%, and even as high as 900 K, 10% of the Bi is diatomic. This may be important for any physical deposition processing of SBT and related compounds for memory films.

Bi_2O_3 layers act as adhesion layers between other metals and dielectrics (including glass). In this respect they are similar to the Ti-adhesion layers used between Pt and oxidized Si substrates.

10. SBT-Family Aurivillius-Phase Layer Structures

The Aurivillius layer-structure compounds come in a family consisting of oxides of general formula

$$A^{2+}M_x^{3+}M_yO_{3z} , \tag{10.1}$$

where A is an alkaline earth (such as Sr), M^{3+} is a trivalent metal (usually Bi), M is a +4 or +5 metal (titanium, tantalum, or niobium), and z can range from 3 to 9. The titanates with even z are usually not ferroelectric along the c-axis; they are antiferroelectric [352, 353], as shown by the superlattice diffraction lines in the HREM data of Scott and Ross in Fig. 10.1. The titanate structures with $z = 4$ and $z = 5$ are shown in Figs. 10.1 and 10.2 and may be compared with that for $z = 2$ in Fig. 9.1. The $z = 5$ ($SrBi_4Ti_4O_{15}$) and $z = 6$ ($Sr_2Bi_4Ti_5O_{18}$) structures are both ferroelectric at room temperature. It is easy to make these at high temperatures from a room-temperature fluorite structure because, as discussed in the preceding chapter, the Ti–O or Ta–O sixfold coordinated octahedra are already formed in that defective structure, the Ta–O or Ti–O bond lengths are very nearly the same in the two phases, and, as shown by Chen and Yu and by Poignant et al. [354], the Bi_2O_2 planes easily slide into the space between adjacent MO_6 octahedra in these structures. Because Bi_2O_2 planes are so easy to position between ABO_3 perovskite blocks ($SrTaO_3$) in SBT, it has been possible to make artificial superlattices by layering the Bi_2O_2 externally, as shown by Tabata et al. [354]. A *functional gradient* device can also be constructed in this way, where the Bi_2O_2 planes are spaced by an amount that slowly decreases or increases with depth z from the electrode interface. Such artificially incommensurate lattices provide a way of tailoring the spontaneous polarization term dP/dz (1.18) to compensate for other (intrinsic) surface effects.

10.1 RBS Studies

Hydrogen contamination is a serious problem in ferroelectric memory devices, including SBT [355, 356, 357, 358, 359]. The most detailed study of the SBT:H problem is that of Zafar et al., who analyzed the hydrogen-reduction of bismuth oxide to bismuth and the related electrode delamination and creation of oxygen vacancies. A more subtle observation was the 22% reduction in grain

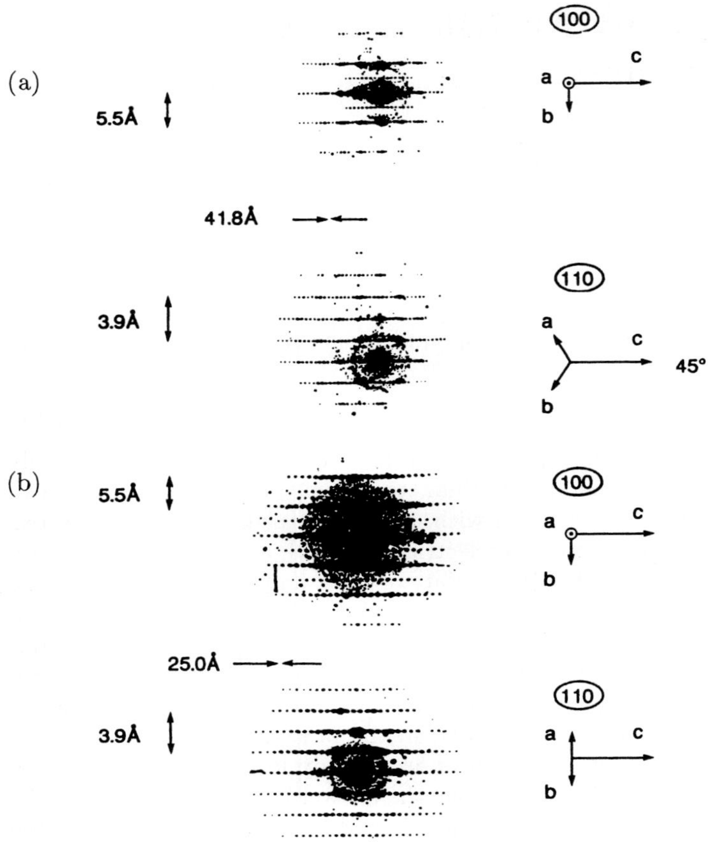

Fig. 10.1. High-resolution electron diffraction data [353] on (**a**) $SrBi_4Ti_4O_{15}$, (**b**) $SrBi_2Ta_2O_9$

size after forming-gas anneal. And Rutherford back-scattering is generally the best technique for measuring hydrogen concentrations and depth profiles in SBT or other ferroelectric memory films [360]. As shown in Fig. 10.3, the hydrogen concentration in SBT is typically 0.05% atomic percent (much less than in BST). Hydrogen contamination in PZT from its forming-gas anneal was studied rather carefully at Fujitsu by Ashida et al. [357] and found to decrease the remanent polarization P_r; and Takatani et al. found [358] that the hydrogen concentration acts as a catalyst to produce oxygen vacancies and positive charge centers [359, 360, 361, 362, 363, 364, 365, 366, 367, 368].

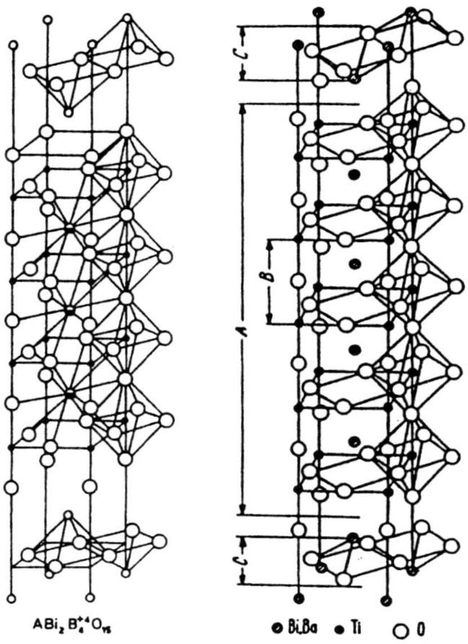

Fig. 10.2. (*Left*) $SrBi_2Ti_4O_{15}$ structure. (*Right*) $Sr_2Bi_4Ti_5O_{18}$ structure

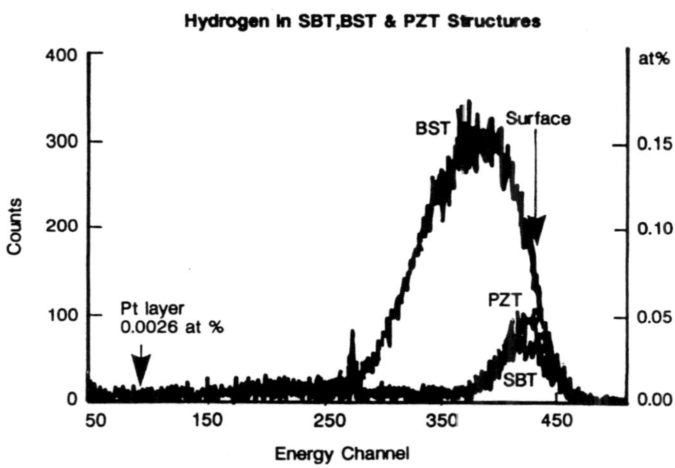

Fig. 10.3. RBS data in BST, PZT, and SBT [356]

10.2 Heavy-Ion Beam Studies

Interdiffusion of ferroelectric and electrode material in ferroelectric memory films is best studied [370] via heavy-ion beam scattering. This technique is quite good for measuring Bi, Sr or Ta diffusion (Fig. 10.4), and can also be used to detect and quantitatively measure residual carbon from MOD processing, as illustrated with the BST data in Fig. 10.5. The carbon comes in lumps, as shown in Fig. 10.6. Similar studies on SBT, illustrated in Fig. 10.7,

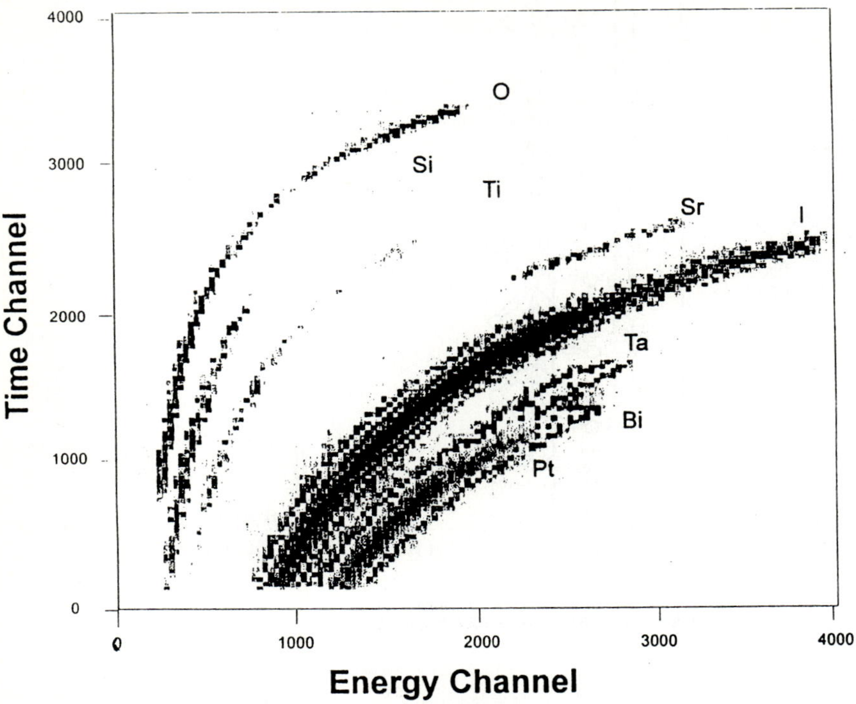

Fig. 10.4. Iodine-127-beam heavy-ion scattering data in SBT/Pt/Ti/Si films, showing depth distribution [356]

showed that unlike the case of PZT (Fig. 10.8), there is no oxygen deficiency in SBT near the electrode interface. Note that the oxygen-deficient region in PZT increases with fatigue, as shown in Fig. 10.9.

10.3 Surface-Science Techniques (XPS, UPS)

More comprehensive studies of the surface and near-surface properties of ferroelectric memory films can be carried out via X-ray Photoelectron Spectroscopy (XPS) and Ultraviolet Photoelectron Spectroscopy (UPS). The basic band structure (Robertson and Chen) of SBT is shown in Fig. 10.10, and its predictions for partial densities of states confirmed experimentally in Fig. 10.11.

A good example of SBT work is the bismuth depth profiles shown in Fig. 10.12a, b. Here we see in the top figure (a film without a top electrode) a dip in Bi concentration at the middle of a film made in two separate spin-on depositions, where the volatile Bi evaporated away. In the bottom figure we see a film with Pt top and bottom electrodes, and we note that the

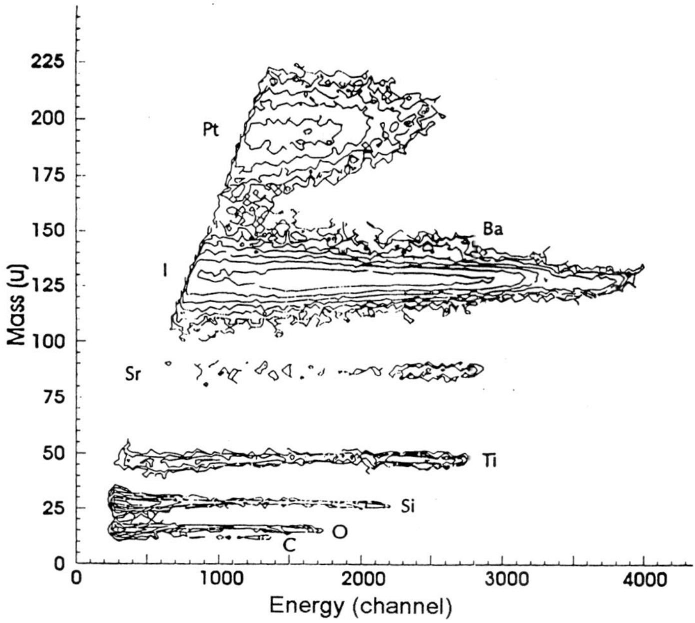

Fig. 10.5. Iodine heavy-ion beam scattering in BST, showing residual carbon at 200 ppm level [356]

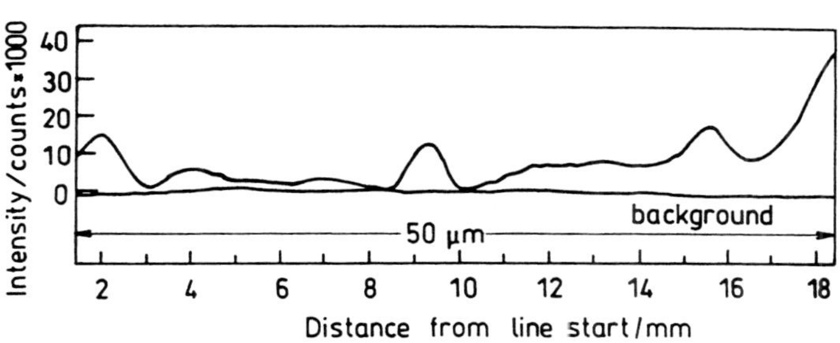

Fig. 10.6. Lateral XPS scan of carbon in BST [356]

bismuth (two peaks near 160 and 164 eV) has percolated all the way through the 200 nm of Pt. Figures 10.13 and 10.14 show that roughly 10% of the Bi remains as elemental metal bismuth; about 90% is Bi_2C_3. At first it is surprising that not all the Bi is oxidized. This could be an artifact of the XPS bombardment but could also be due to interstitial hydrogen in the SBT.

XPS is particularly good at measuring the electron affinities and band matchups in ferroelectric semiconductors. The role of excess Bi is important in determining these matchups. In contrast to Fig. 5.20, which showed a Schottky barrier of about 1 eV for SBT/Pt, Fig. 10.15 shows [371] that excess bismuth forms a surface trap that pins the Pt electrode work function

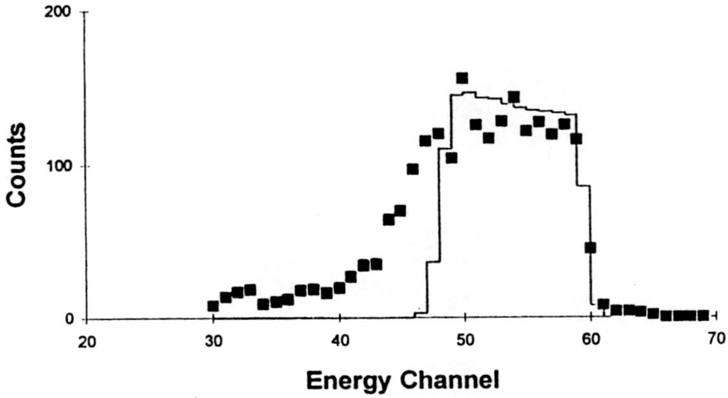

Fig. 10.7. Oxygen stoichiometry profile in SBT/Pt. The *solid curve* shows the top edge (at 60 units) and the bottom edge (at 47 units) of the SBT and demonstrates that there is no oxygen depletion at the bottom electrode

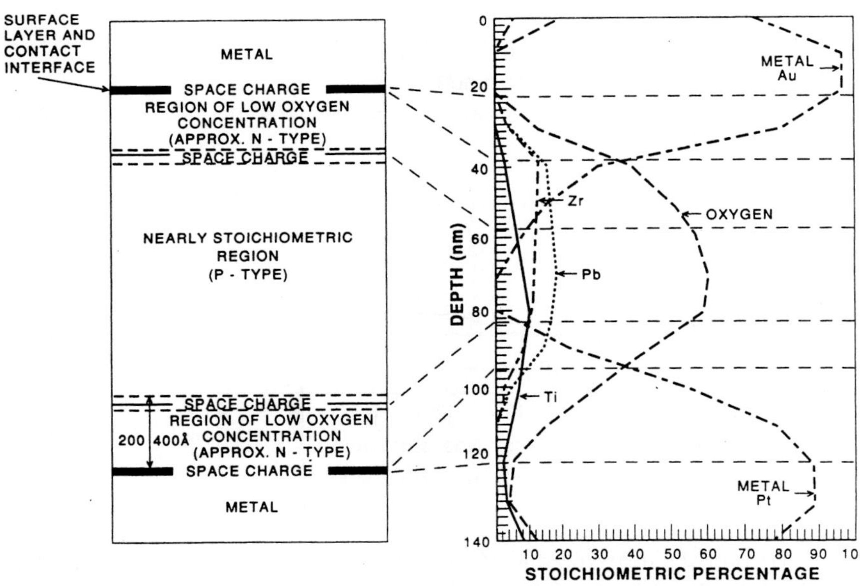

Fig. 10.8. Depth profile of PZT, showing oxygen depletion near top and bottom electrodes

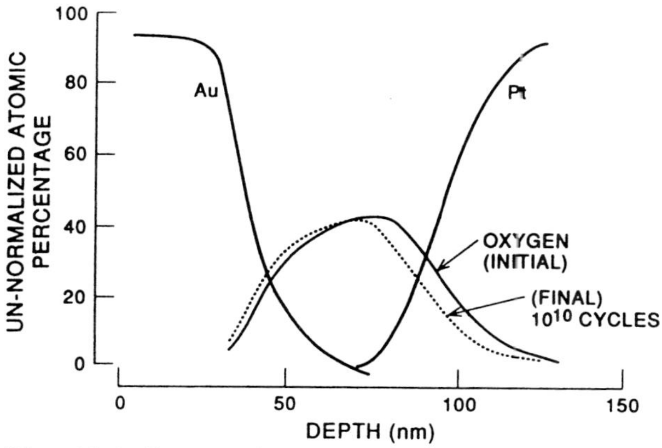

Fig. 10.9. Increase in oxygen depletion near electrode in PZT/Pt after fatigue [369]

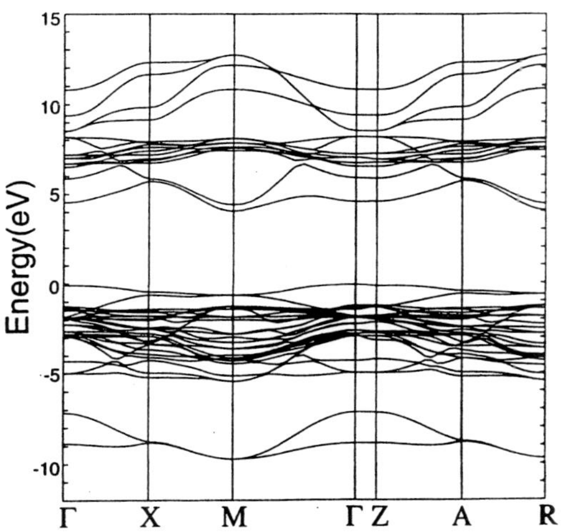

Fig. 10.10. Band structure of SBT [360]

so that the resulting junction is quasi-ohmic. And Fig. 10.16 confirms via surface photovoltaic measurements that this Bi trap lies at approximately 2.3 eV into the bandgap in SBT [372].

The trap energies for several common impurities in ferroelectric memory films are well established. For example, the V_{Ti} (titanium vacancy) level in PZT is 0.5 eV into the gap, and Pb^{+3} is at 0.3 eV [373]. The oxygen vacancy V_0 is positively charged and is a shallow donor [374], with migration enthalpy about 1.0 eV. These are probably the three most important electronic traps in PZT. For lead titanate (PT) Pintilie et al. find [375] via Thermally Stimulated Current (TSC) techniques that there is only one main trap, at 0.30 eV

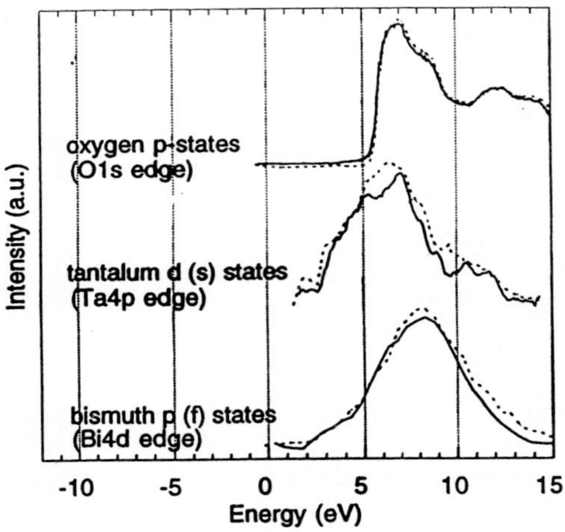

Fig. 10.11. Partial densities of states for SBT [360]. *Solid curve* is photon yield and *dashed curve* is electron yield

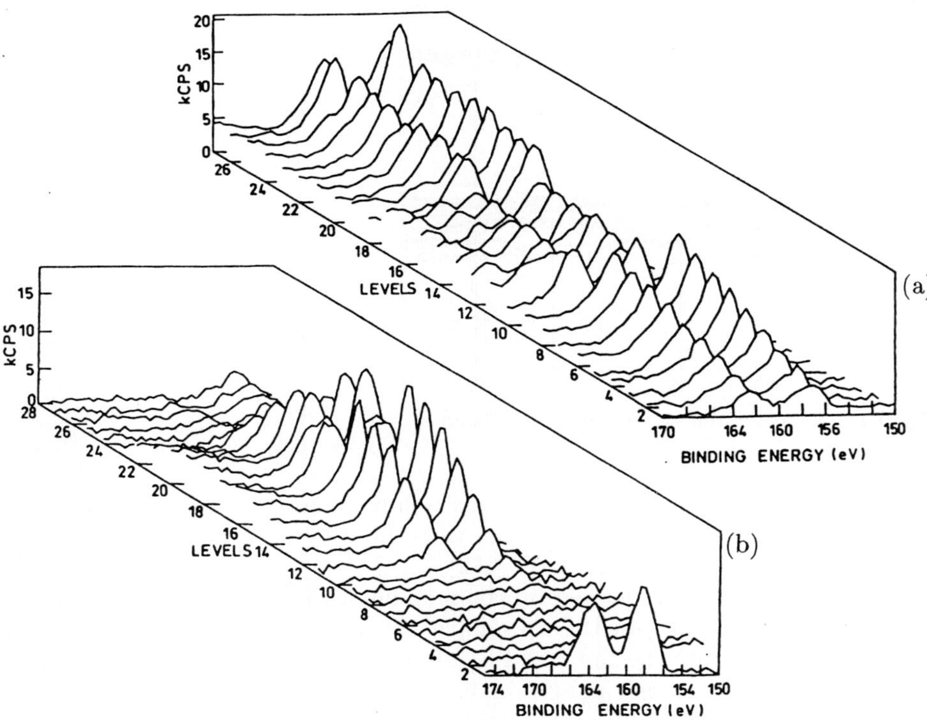

Fig. 10.12. XPS depth profiles of Bi in SBT films (**a**) without electrodes, (**b**) with Pt top and bottom electrodes

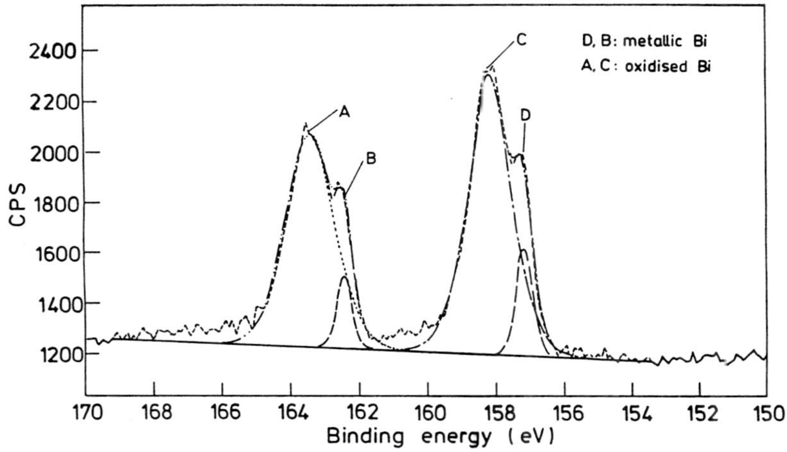

Fig. 10.13. Resolution of Bi peaks in Fig. 10.12 into metallic Bi and Bi^{+3}

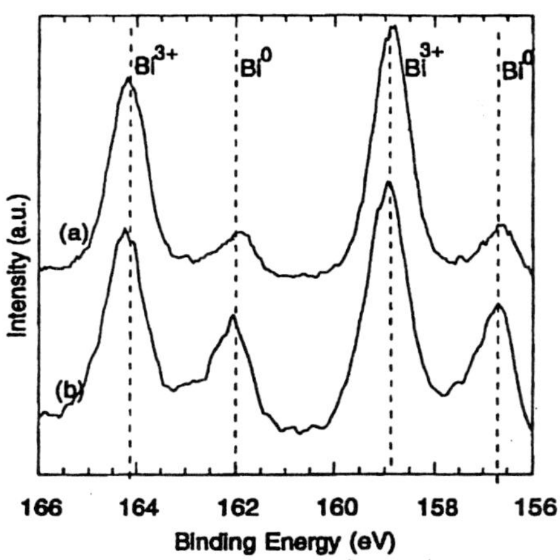

Fig. 10.14. Angular effects showing that the neutral Bi is primarily at the surface. *Top trace*, normal incidence; *bottom trace*, grazing incidence to enhance surface contributions

in samples annealed at 923 K. However, without such an anneal, three other traps exist at 0.23, 0.34, and 0.40 eV. This shows the utility of annealing out traps. For $BaTiO_3$ the early work of Kudzin [376] on the temperature dependence of fatigue revealed (Fig. 10.17) a single trap at 0.07 eV dominating behaviour. This was probably the first study in which the energy of a trap in a ferroelectric was extracted from fatigue data at different temperatures. In $Bi_4Ti_3O_{12}$ the dominant traps are at 0.55 and 0.60 eV [377].

Photoelectric effects (both photovoltaic and photoconducting) are particularly useful for the determination of trap energies in ferroelectric thin films.

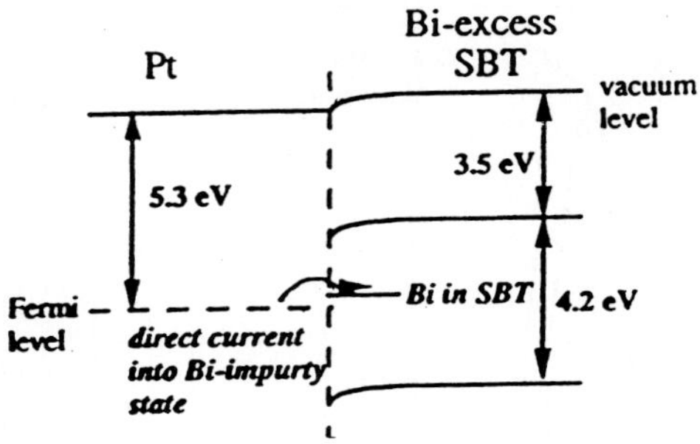

Fig. 10.15. Band structure matchup for SBT:Bi (SBT with excess bismuth) on Pt electrodes [359]

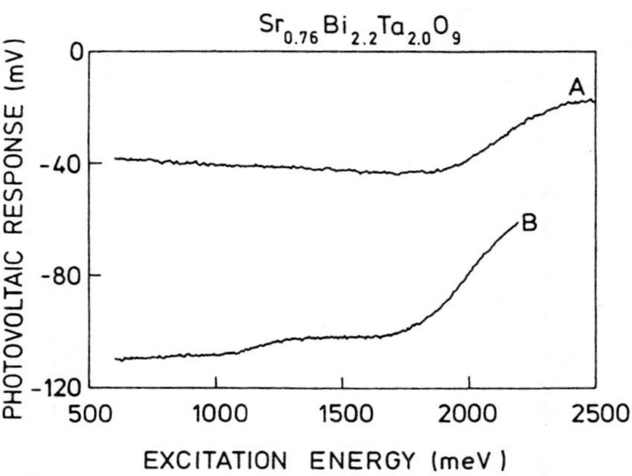

Fig. 10.16. *Curve A*: Surface photovoltaic measurements, showing very large density of states near 2.2 eV due to excess bismuth [360]. *Curve B*: Second sample

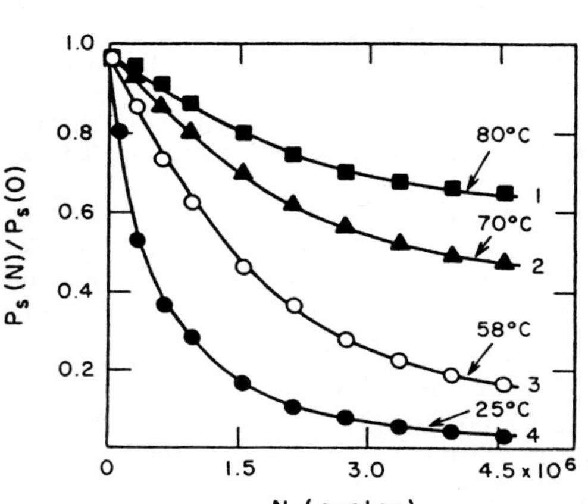

Fig. 10.17. Fatigue of barium titanate at different temperatures [376]

Generally the photocurrent peaks at the band gap and the photovoltage at the dominant trap energies, but in systems with high donor densities impurity banding can give photocurrents at the trap energies. Studies of bismuth titanate [377] and lead titanate [378] have been reported recently, but the $PbTiO_3$/Si data were dominated by band-to-band transitions at 0.38 µm ($PbTiO_3$) and 0.9 µm (Si) and not trap transitions. More interesting is the latest work on SBT photoconductivity [379], which confirms the bandgap at 3.94 eV [380].

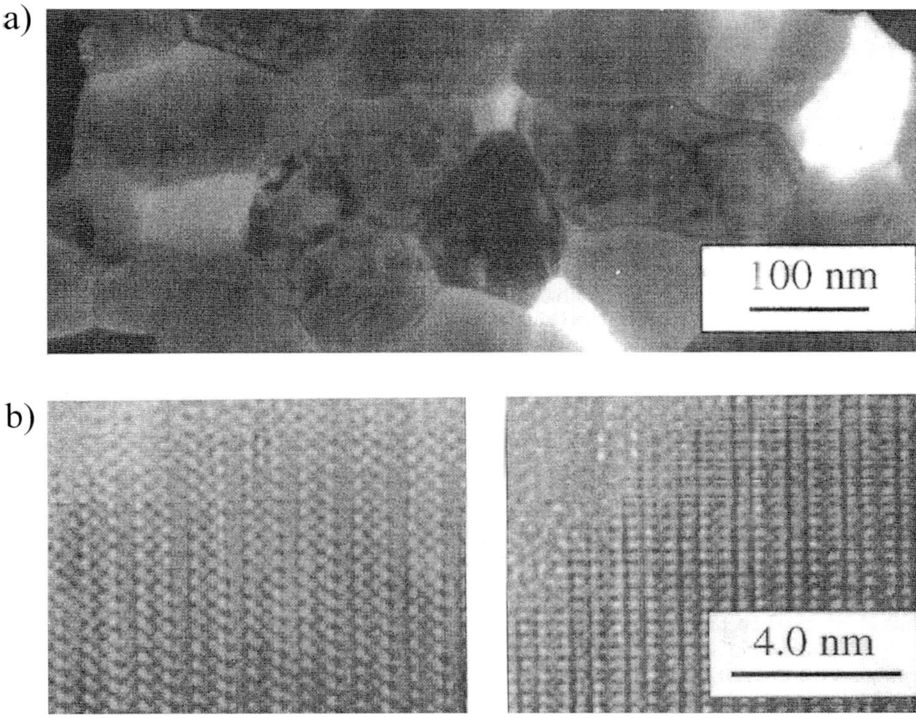

Fig. 10.18. (a) *Top*: globular polyhedral grains in SBT. (b) *Bottom*: atomic resolution electron micrographs of SBT

The microstructures of SBT differ considerably from those in PZT. PZT normally crystallizes in thin film form in fine grained ceramics (40 nm) that are grain-oriented in the form of columnar grains that extend all the way from top to bottom electrodes. In contrast, SBT usually grows as globular polyhedra (Fig. 10.18).

11. Deposition and Processing

A variety of deposition techniques are employed for ferroelectric thin film capacitors to be used in memory devices. These include sputtering [382], MOCVD (Metal-Organic Chemical Vapour Deposition) [383], MOD (Metal-Organic Decomposition) [384], PLD (Pulsed Laser Deposition) [385], sol–gel (solution–gelation) spin-on techniques [386], and a *mist* deposition in which stoichiometrically correct submicron droplets are sprayed into the deposition chamber [387]. The MOCVD techniques include a *flash* process [388, 389, 390, 391, 392] in which the precursor solutions are delivered to the deposition chamber in liquid form (due to their low volatilities) and then flashed into the vapour phase via incandescent lights or laser sources at a shower head delivery point just below the wafer. These flash systems all differ from each other (Isobe's achieves the highest deposition rate), and all have separate patents. What they share is the ability to handle viscous precursors with low vapour pressures. [The LG Semiconductor Corp. flash process for BST utilizes relatively viscous barium- and strontium-$(THD)_2$-tetraglyme, together with Ti(O-iso-Pr)$(THD)_2$.] True CVD is not used; that is, vapour deposition with chloride or hydride precursors, since halogen contamination is a problem. MBE (Molecular Beam Epitaxy) is generally not used, with one or two research exceptions [393]. And thick film techniques such as spray pyrolysis, slip-casting, etc., are not used. Thermal evaporation in a Pt crucible can be used for $BaMgF_4$ [394].

There are three general ways of preparing oxides via decomposition of metal alkoxides. Taking zirconium as an example [395]:

$$Zr(OR)_4 = ZrO_2 + C_nH_{2n} + R_2O + ROH \,, \quad (11.1)$$

$$Zr(OR)_4 = ZrO_2 + CO_2 + H_2O \,, \quad (11.2)$$

$$Zr(OR)_4 = ZrO_2 \times nH_2O = ZrO_2 + H_2O \,. \quad (11.3)$$

Relations (11.1) and (11.2) yield high-purity oxides and only gaseous by-products and are a kind of pyrolysis. But (11.3) is the hydrolytic decomposition of an alkoxide (usually in an organic solvent) via precipitation of hydroxides. These hydroxides form oxides on dehydration, and that is the basis of most present sol–gel ferroelectric oxide technology for thin-film memory capacitors.

When BST, SBT, or PZT is processed via any of the techniques below, there are a number of choices for its electrodes. NEC uses RuO_2, Mitsubishi employs elemental Ru, and Samsung and the US DRAM consortium both utilize Pt.

Independent of the choice of deposition technique, there remains the separate problem of etching. In general, selective wet-etch techniques do not exist for any of these materials. Dry etching has explored various methods [395], of which RIE (Reactive Ion Etching) and ECR (Electron Cyclotron Resonance techniques) are leading. These give rates of 10–50 nm/min with a selectivity as great as 5:1 for PZT compared with Pt. Optical patterning has been exploited by Mitsubishi–Symetrix, by adding a photoresist to the ferroelectric sol–gel precursors. More recently this technique was also employed at Thomson CSF [395].

11.1 Sol–Gel Spin-on

The use of solution–gelation techniques for preparation of ferroelectric thin films of the PZT family was pioneered by Turova and Yanovskaya as part of the latter's PhD thesis in Moscow in 1973 [396]. Sol–gel PZT was patented in the USA in 1975 [396]. There are many precursors available, but a typical process employs titanium iso-propoxide, zirconium propoxide and lead acetate. Butoxides and other alkoxides may be used as well. The extension to very complicated alkoxide precursors was recently described by Kato [397]. It is usual to make up (and refrigerate) stock solutions. The first stock solution is of the stoichiometrically correct mix of Pb, Zr, and Ti precursors in the case of PZT; and the second solution is the hydrolysis solution. Typically 50 cm^3 of hydrolysis solution with 0.2 moles of distilled, deionized water and 0.03 moles of NaOH are prepared as a stock solution in 2-methoxyethanol. This solution is later hydrolyzed by mixing a molar stock solution with a molar stock solution of Pb/Zr/Ti alkoxide precursor in a 2:1 ratio [398]. The PZT precursor solutions vary. For sol–gel SBT bismuth acetate and tantalum ethoxide work well. The solution is aged overnight. Methoxyethanol is then added, sometimes with a drying agent such as formamide. Independent of the precursors, the resulting solution is rather non-viscous and can be dripped onto a photoresist spinner at around 1500–2000 rpm for about 1 min. This produces a film about 50–100 nm thick. Thicker films (200–300 nm) require repeating the process three to five times. Usually the film is annealed between each layer deposition.

A common technique for the general preparation of the alkoxide metals used in the above stock solution is to react the elemental metal with alcohol. In order to make this reaction go to completion it is necessary to supply 2.3–3.0 V to the system (often in an electrolytic cell), and to dissolve the alkoxides obtained (with excess alcohol) at elevated temperatures. Sometimes it is desirable to react the metal without an excess of alcohol but in an inert solvent (usually xylene).

A second technique is direct anodic oxidation of the metal. This is useful for less active metals, such as In, which dissolve very slowly. Zr and Ti for PZT can both be prepared by anodic dissolution in alcohol, facilitated by small amounts of ammonium chloride. The general advantage of this technique is that it starts with elemental metals, which are often available as ultra-high purity powders.

A third technique is the reaction of oxides or hydrides with alcohols. This method can be used with thallium ethoxide or vanadium systems. Many other techniques exist for virtually all metals, including alcoholysis of MX_n (where X is H, NH_2, SH, etc.), esterification, exchange reactions of metal halogenides with alkaline alkoxides, and redox reactions.

11.2 Sputtering

Plasma sputtering of most ferroelectric oxides [399, 400, 401, 402, 403] has been successful using either elemental or alloy targets for BST, PZT, PLT, PLZT, and the basic barium and lead titanate compounds. Typically this requires low deposition rates (< 2 nm/min), gas pressures of 1–10 mtorr, and for the lead compounds, targets rich in Pb or PbO (to compensate for high volatility) [399].

Sputtering of ferroelectric oxide films can be carried out via diode sputtering or magnetron sputtering. Rather good leakage current performance is usually exhibited by these sputtered films: typically 10 nA/cm^2 across 180 nm of PZT on oxidized Si [404] with a switched charge of 4.7 fF/μm^2 for NV-FeRAMs, and 12.6 fF/μm^2 total charge density on strontium titanate DRAM films, equivalent to a SiO_2 thickness of 2.7 nm. SBT has been successfully sputtered at NEC and elsewhere [404, 405, 406, 407, 408, 409]. BST also sputters well.

Yang et al. achieved a particularly good remanent polarization of $2P_r =$ 19 μC/cm^2 in rf magnetron sputtered SBT [409], but with an unacceptably large coercive field of 150 kV/cm.

A rather special extension of sputtering techniques for ferroelectric oxide films is Multi-Ion-Beam Reactive Sputtering (MIBERS). This technique attempts to minimize certain traditional sputtering problems, such as the large differences in sputter yield for different elements in the target, and the presence of negative ions and reflected neutral atoms, which cause nonstoichiometry, surface damage, and uncontrolled local film growth. MIBERS sputters different elements independently and simultaneously via different ion beams. It has achieved rather good success with PZT. Its main disadvantages are that the equipment is expensive and rather complicated to run.

Recent sputtering work has emphasized epitaxial growth as well as coverage of large areas. It is possible to grow epitaxial (111) PZT or PLZT on a variety of substrates, including sapphire. Sputtering has the advantage that it is a mature technology and affords good step-coverage for high-density ICs. Its great disadvantage is that it is not spatially uniform over a large area wafer (8″ or 12″) and it exhibits target aging and hence a lack of uniformity from wafer to wafer. In my opinion it is not the preferred technique for commercial ferroelectric passgate (1T–1C) memories, but it might find a use in FE-FETs, where rf magnetron sputtering has been used for making PZT, PT, and BMF gates [410].

11.3 Metal-Organic Chemical Vapour Deposition (MOCVD)

The use of MOCVD [411] for PZT was patented by Burn in 1978 [412]. By 1991 one could achieve good device-quality films [413, 414]. For SBT the use of MOCVD is more difficult because the precursors used are more viscous and have lower vapour pressure. For practical devices and commercialization, it is desirable to have vapour pressures > 50 mtorr at ambient temperatures. To overcome this problem one utilizes a delivery system independently developed by Isobe at SONY Corp. and Desu (now at U. Mass.) in which the precursors are brought to the deposition chamber in liquid phase, then abruptly *flashed* into vapour with a high-intensity lamp. Actual deposition is from a showerhead device located just below the wafer. Alternatively, one can dissolve the precursors in a liquid carrier, such as tetrahydrofuran [415], hexane, or monoglyme.

MOCVD of BST also works well [416, 417, 418, 419, 420] and has perhaps been most successfully commercialized at Advanced Technology Materials Corp. in the USA [403] and by Siemens in Germany [403]. It often employs $TiO\text{-}(THD)_2$, where THD is 2,2,6,6-tetramethyl-3,5-heptadione. BST MOCVD often uses instead 'ack-ack' – strontium-hexafluoroacetylacetonate-tetraglyme [404]. IBM has reported an ack-ack MOD route for BST that utilizes barium and strontium acetate [404]. A disadvantage of this route is a larger thermal budget and higher carbonate residual impurities. Lindner et al. have had recent success with a mixed Ba/Ti precursor – $Ba_2Ti_2(thd)_4(OEt)_8(EtOH)_2$ – originally developed by Kessler et al. [421].

Very recently Kaiser et al. [422] have shown that the grain orientation of MOCVD BST films (*a*-axis normal, *c*-axis normal, or mixed) depends strongly on the Sr concentration (MgO substrates). This is due to the formation of a thin layer of amorphous BST and to the fact that the amorphous layer exhibits a 1% compressional strain. The BST amorphous layer was independently verified by Dedyk et al. [422]. In a similar vein, the role of Nb in SBTN films in determining their grain orientations, first pointed out as an empirical observation by Atsuki et al. [422] and by Noguchi et al. [422], was explained by Watanabe et al. [422].

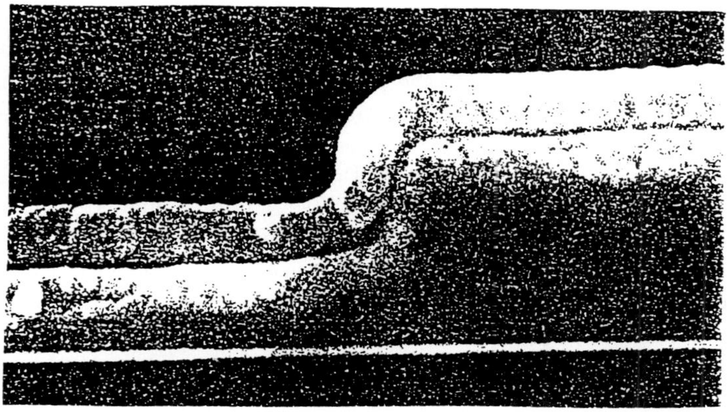

Fig. 11.1. Step coverage of BST on Pt, showing a vertical step of 500 nm in Si, covered with 200 nm of Pt, 40 nm of mist BST, and a top 200 nm of Pt. McMillan L. D., private communication cited in [426]

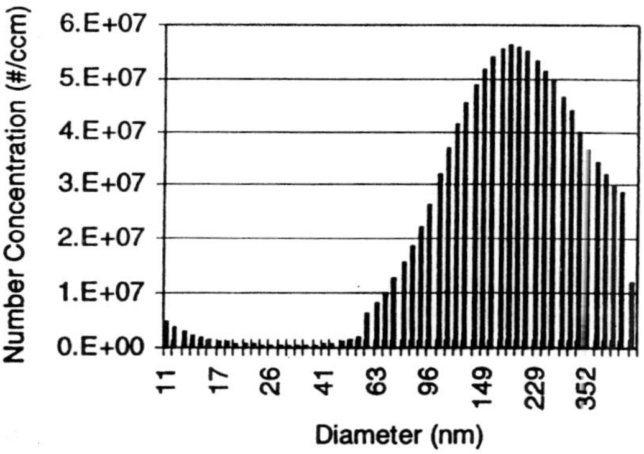

Fig. 11.2. Particle size distribution from the mist deposition system for SBT (octane-based MOD solution) [423]

Only peripherally related to CVD deposition is the so-called *mist* system [423, 424, 425, 426]. Sometimes termed liquid CVD this technique injects submicron-diameter droplets of stoichiometrically correct precursors via an aerosol nozzle driven ultrasonically at around 1.7 MHz into the sample chamber. Typical parameters are 2 liters/min gas flow, deposition pressure slightly (50 torr) under ambient, 15 rpm wafer rotation, in a 5 kV dc voltage with charged particulate deposition (5e). Charging of the wafer is prevented by flowing a neutralizer gas at around 50 sccm. As shown in Fig. 11.1 [426], it gives remarkably good step coverage and high deposition rates (> 18 nm/min). This system is available commercially from Samco Corp. in Japan and from SubMicron Systems Corp. in Pennsylvania. Figure 11.2 shows the particle size distribution typically obtained with the mist deposition system, using an octane-based SBT MOD solution; the distribution peaks at 0.21 μm [423]. These sizes were determined by mass separation techniques, using electrical charging of the particles ($q \approx 5e$ each). It would be very interesting to measure this size distribution independently via Mie scattering, which is ideal for 0.2 micron particles using a He–Ne laser.

MOCVD has an added advantage for PZT work in that it can be used to deposit RuO_2 electrodes in addition to the PZT. A good review of RuO_2 electrodes is given by Bai et al. [427]. RuO_2 electrodes for BST are in my opinion too leaky for real commercial DRAM devices, but leakage current is less important for PZT NV-FeRAMs. Other authors believe that RuO_2 bottom electrodes will be very good for BST DRAMs [428, 429, 430]; the ruthenium oxide can be epitaxial, and such a BST device has recently been described in some detail.

Plasma enchanced MOCVD of ferroelectric films has been well reviewed by Dey and Alluri [430]. PE-MOCVD has an advantage over thermal MOCVD in that it can be used with materials that require low deposition temperatures. The interfaces and doping profiles of previously deposited layers are left relatively undisturbed in this way, and solid state diffusion is minimized. Using PE-CVD deposition temperatures for Aurivillius-phase SBT as low as 550°C have been claimed. However, I remain skeptical of this temperature, as it is difficult to ascertain the true surface temperature with PE-CVD techniques.

MOCVD processes can be one-step or two-step. Two-step processes for PZT seem to be superior with respect to surface roughness. In the Emcore two-step process a thin nucleation layer is deposited at a relatively low temperature, and then the deposition temperature is increased for rapid grain growth. The reactor wall temperature T is controllable (423 K $< T <$ 523 K), and the precursors used (lead and zirconium thd_2) do not decompose until higher temperatures (600 K). In order to be commercially feasible, high deposition rates must be obtainable with any MOCVD process. At present, deposition rates vary from a low of about 0.1 microns/hr to a high of about 15 microns/hr. If the deposition rates are too high, the surface mobility is insufficient to produce good films (in PZT this results in pyrochlore).

Photo-enhanced MOCVD has been used for these materials [432]. Typically a 500 W xenon–mercury lamp is employed. This has produced some surprising results, e.g., that (111) $PbTiO_3$ could be deposited directly on Al_2O_3 only under illumination. Although illumination in this way can reduce the minimum deposition temperature (not surprising) it does not always increase the deposition rate; deposition rates actually decrease for lead titanate (surprising).

11.4 Pulsed Laser Deposition (PLD)

Laser ablation of PZT [431, 432] was invented by Maher [433] in 1979, but early efforts were plagued with particulate contamination and surface roughness. Roy et al. achieved an early result [434] on 250 nm thick $SrTiO_3$ DRAM films of 8.4 fF/μm^2 with extremely low leakage currents of 2 nA/cm^2. Ramesh [435] and later Lin et al. [436] prepared excellent PZT via PLD on YBCO superconducting electrodes with low coercive fields but only fair remanent polarization (6 μC/cm^2). Later this group reported good results with PZT on yttrium-stabilized zirconia [437]. A typical PLD deposition of ferroelectric thin films for memory applications utilizes KrF excimer lasers at a wavelength of 248 nm and a repetition rate of 5 Hz. Satisfactory deposition rates and film quality are achieved with power densities of 2 J/cm^2. In order to obtain uniformity, a rotating target and oblique angle of incidence are usually employed. Rather good PLD films of Au/PZT/$SrRuO_3$ have been reported recently from Hitachi [438]. But despite excellent remanent polarization values and good fatigue performance, these exhibit poor leakage current properties and relatively high ($E_c = 90$ kV/cm) coercive fields. Both the high E_c values and apparent 'steps' in the hysteresis curves presumably arise from damaged surface layers, which are often a problem with PZT.

Pignolet et al. have been particularly successful in achieving large-area (about 20 cm^2) epitaxy of ferroelectric oxides [439]. Rather remarkably, they have also achieved a coercive field of only 20 kV/cm (2 MV/m) in PLD *c*-axis epitaxial SBT, about 1/3 the normal value. Such a field would permit 1.1 V operation at saturated polarizations (about $3E_c$) at film thicknesses of 200 nm, which would make a high-yield commercial product.

For both SBT and isomorphic barium bismuth titanate PLD produces via tilted targets (Fig. 11.3) large-area, uniform thickness (Fig. 11.4) textured films which are mostly (001) *c*-oriented, as shown in Fig. 11.5, but having rectangular (110) crystallites protruding out of the surface. In contrast to this, the regions without these rectangular crystallites are exceptionally smooth, with an rms roughness of only 2.5 nm over a 6 μm^2 area. Figure 11.6 shows a high-resolution TEM cross section of SBT films in the region in which these rectangular (110) crystallites are found. The vertical lines show the Bi_2O_2 planes separated by 1.25 nm (half the 2.5026 nm unit cell lattice constant). By preparing such films one can measure the *c*-axis switching in SBT (Fig. 11.7), which has hysteresis that is nearly square, and also the response perpendicular to the *c*-axis (SBT II and III in Fig. 11.7), which is fully saturated at 4 V.

Recently Matsumuro et al. have reported growth of (105)-oriented SBT via PLD at temperatures as low as 350°C [439].

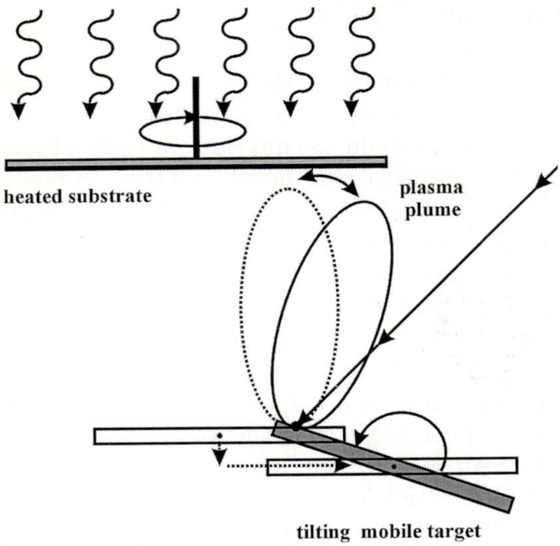

Fig. 11.3. Geometry of tilted target PLD

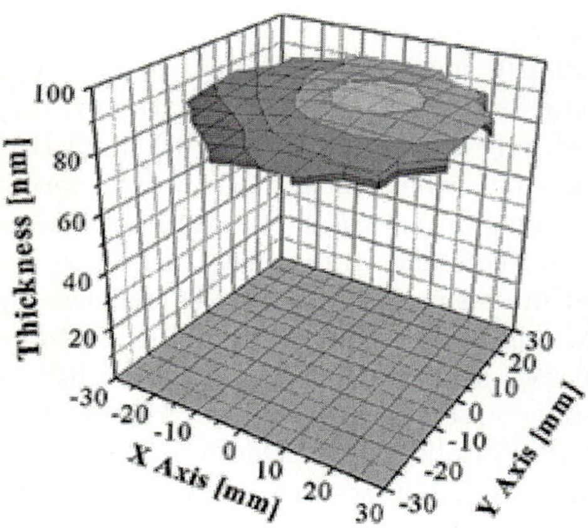

Fig. 11.4. Uniform large-area thickness of PLD SBT films

11.5 Metal-Organic Decomposition (MOD)

Sol–gel and MOD both use the decomposition of metal alkoxides to produce oxides. However, the sol–gel technique strictly refers to solutions (in alcohol) which undergo gel formation, whereas MOD does not. Neither 'sol–gel' nor 'MOD' is the term of choice now for reference to these techniques. Together with MOCVD, they are usually lumped together as CSD – Chemical Solution Deposition – a term that encompasses almost all wet-chemistry techniques (as opposed to physical deposition, such as sputtering or MBE).

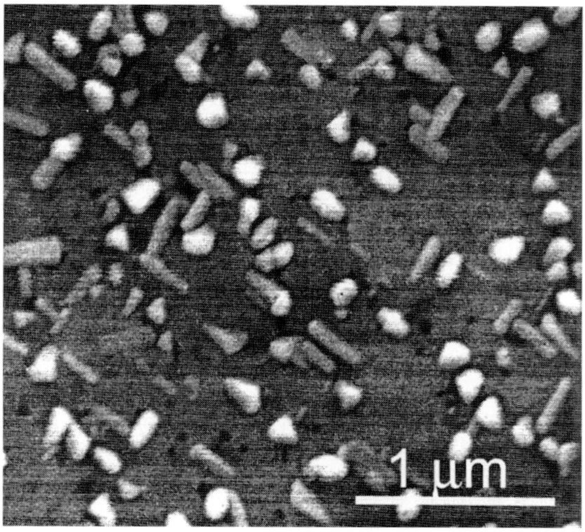

Fig. 11.5. Scanning force microscopy topographic image of epitaxial PLD barium bismuth titanate film, showing (110) crystallites protruding out of the (001) flat surface [439]

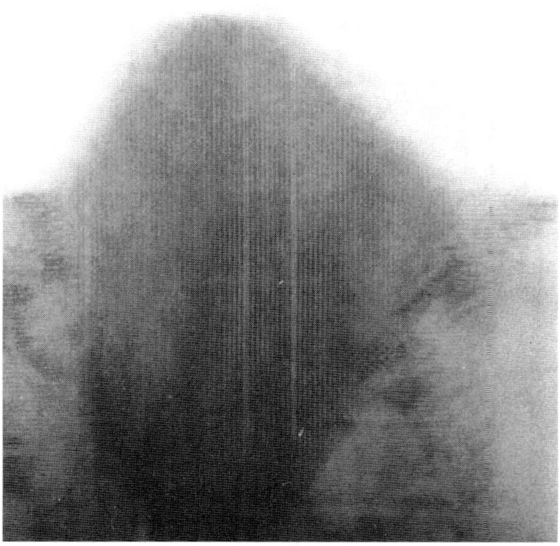

Fig. 11.6. TEM cross-sectional micrograph of SBT (110) grains embedded in a flat (001) film [439]

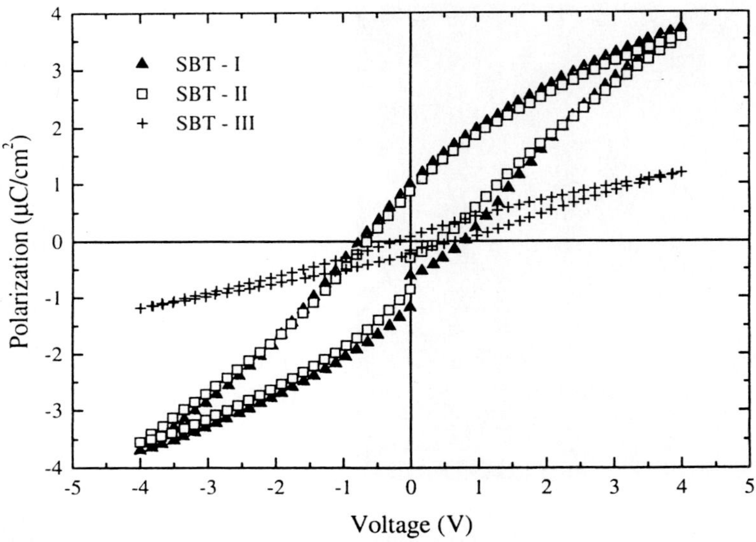

Fig. 11.7. Switching response in SBT along the *c*-axis (+) and perpendicular to the *c*-axis (□ and ▲)

MOD is a popular technique for deposition of BST [440], where Matsushita has reported a capacitance of 32 fF/μm^2 on a 140 nm film with excellent leakage current (10 nA/cm^2 at the intended working voltage of 3.3 V) [441].

11.6 Molecular Beam Epitaxy (MBE)

MBE has been used in a very limited way to carry out research on ferroelectric thin films for memory device applications. Gutleben [442] reported results on SBT/Pt films as a function of thickness, using a small MBE system built in to his Scientia ESCA apparatus. This permitted some rather subtle variations to be characterized. More recently McKee et al. [443] have produced extraordinarily good (defect-free) $SrTiO_3$/Si interfaces via MBE. Since these are pure strontium titanate, they are not ferroelectric and not usable as ferroelectric gates in FETs. However, the technique is extremely promising for nondestructive-read-out memories.

12. Nondestructive Read-Out Devices

All the devices discussed so far in this book have destructive READ operations. Since a memory may be written perhaps only 10^6 times but read 10^{12} times, this reset operation involves a lot of switching and fatigue that could be eliminated in a NonDestructive Read-Out device (NDRO).

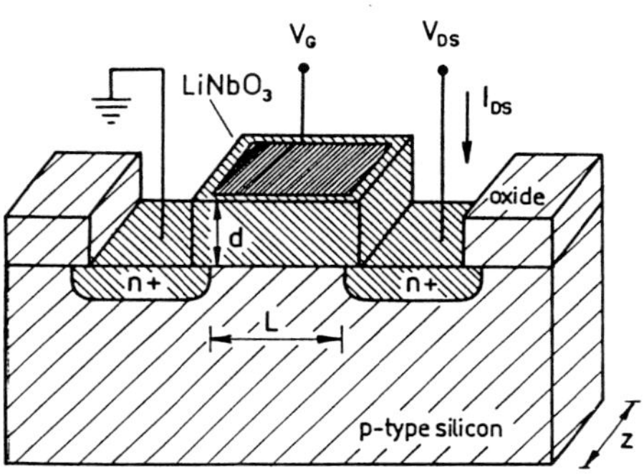

Fig. 12.1. Ferroelectric gate field-effect transistor (FE-FET) [444]

12.1 Ferroelectric Field-Effect Transistors (FETs)

The NDRO device shown [444, 445] in Fig. 12.1 has a ferroelectric film replacing the metal gate in an FET. In this geometry, polarizing the film + or − changes the source–drain current. Thus, by monitoring the magnitude of the source–drain current, one can read the 1 or 0 in the bit without ever switching or resetting the ferroelectric polarization. The main difficulty with this device structure is that there must be a direct ferroelectric–silicon interface. Since the surface trap density of ferroelectric oxides is very large (about 10^{20} cm^{-3}), this causes considerable charge injection from emptying traps, and a degradation in the switched charge measurable due to the displacement vector D in the switching capacitor.

Figure 12.3 shows the charge motion in a ferroelectric FET during its operation [446, 447, 448]. This causes the $P(E)$ hysteresis to be swept out

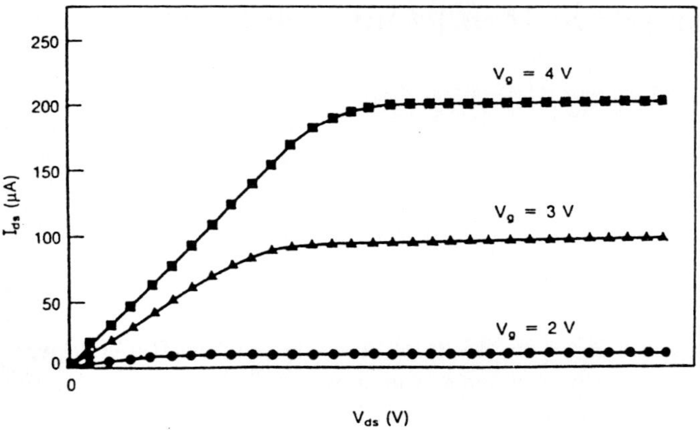

Fig. 12.2. Source–drain current versus source–drain voltage at different gate voltages

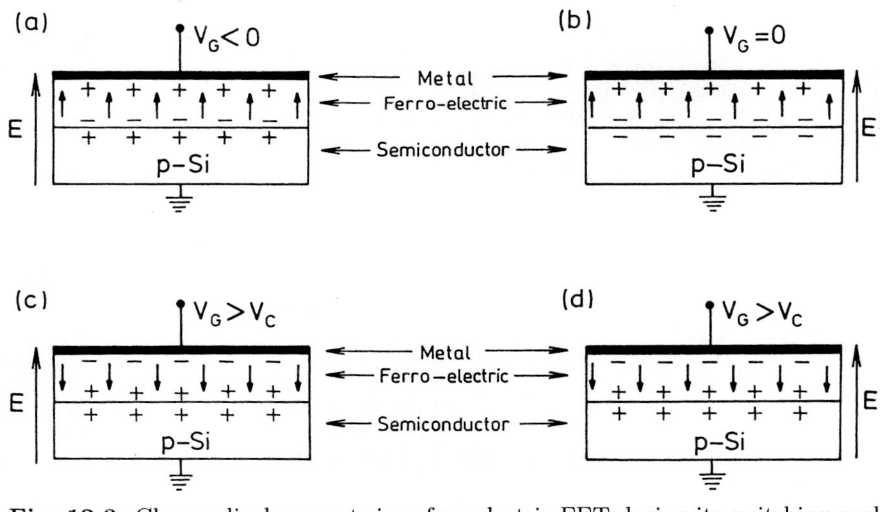

Fig. 12.3. Charge displacements in a ferroelectric FET during its switching cycle [446, 447, 448]

(Fig. 12.4) in a path which is clockwise [448] as opposed to the counter-clockwise path traced out by charge injection devices. Thus the direction of this hysteresis path can be used to determine whether the device is exhibiting true ferroelectric hysteresis or not.

Early embodiments of ferroelectric FETs included the lithium niobate device from Rice University shown in Fig. 12.1, and devices [449, 450, 451] using $BaMgF_4$, illustrated [452] in Figs. 12.5 and 12.6. The $LiNbO_3$ device obtained a gate current of 34 pA at 1.0 V with a large gate area of 625 μm^2.

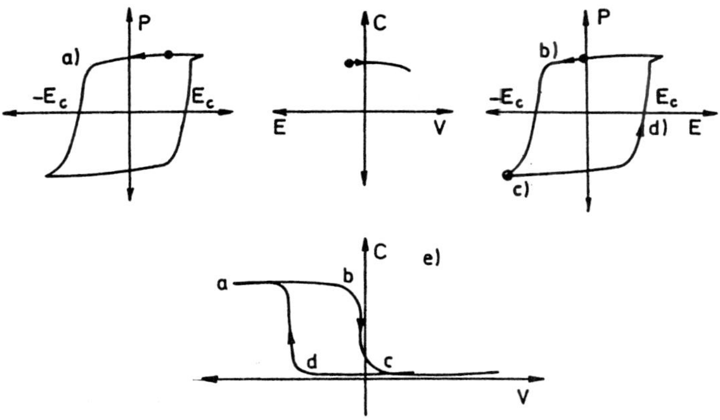

Fig. 12.4. Electrical hysteresis associated with the switching cycle in Fig. 12.3 for an FE-FET [446, 447, 448]

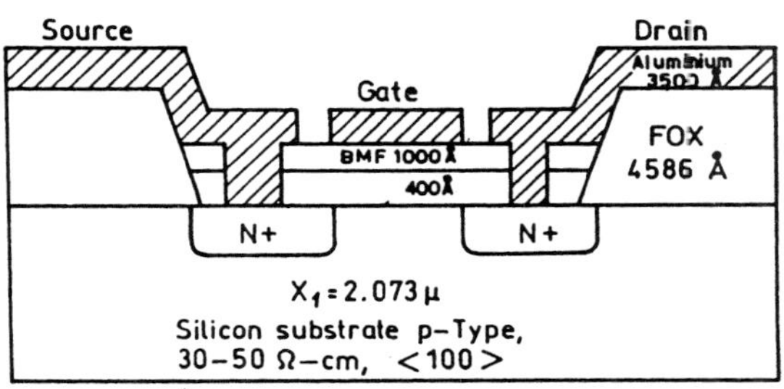

Fig. 12.5. $BaMgF_4$ FE-FET (schematic): buffer is 40 nm of $Si0_2$. Symetrix [452]

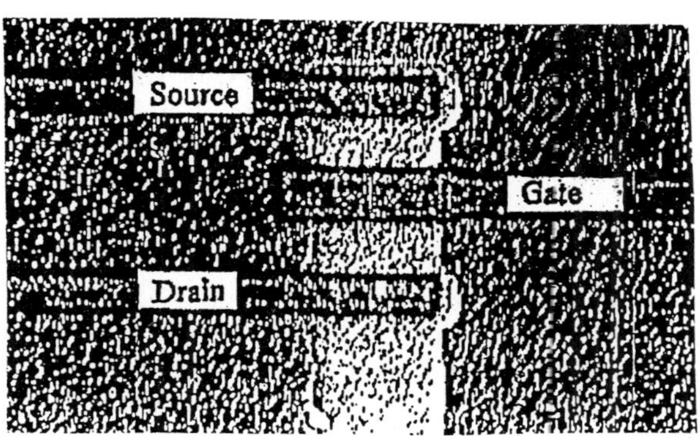

Fig. 12.6. $BaMgF_4$ FE-FET (microphotograph) [452]

Figure 12.2 shows its source–drain current versus source–drain voltage at gate voltages of 2, 3 and 4 V.

Parenthetically, in relation to the discussion of space-charge-limited currents in Chap. 4, we note that lead titanate gated FETs at fields greater than 35 MV/m clearly exhibit SCLC [453]. Autran et al. [454] have independently verified SCLC in PZT films.

More recent FE-FETs have employed PLZT gates [455] or have been designed for SOI (Semiconductor-On-Insulator) applications [456, 457], using a variety of different ferroelectrics [458].

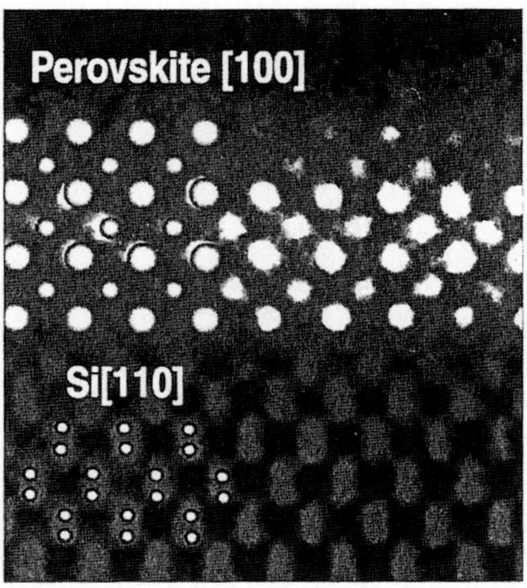

Fig. 12.7. First five monolayers of $SrTiO_3$ deposited directly on Si via MBE [459]

A breakthrough was made in 1998 by McKee et al. at Oak Ridge [459]. They succeeded (Fig. 12.7) in growing a high quality layer of strontium titanate [100] on a clean silicon surface without creating defects at the interface [460]. They used MBE to put 15 nm of $SrTiO_3$ directly onto Si. The key step or trick in their process was to first clean the Si in ultrahigh vacuum and then to grow a single atomic layer of strontium silicide on it at 1123 K. The $SrTiO_3$ is then grown via MBE of alternating atomic layers of SrO and TiO_2. The resulting interfacial trap density is 10^{10}–10^{11} cm^{-2}, about 100 times lower than in conventional BST/Pt devices and as good as that in any conventional gate oxide (SiO_2 or ONO) currently used in silicon microelectronics. (The lowest trap density obtained for chemical solution deposition of bismuth titanate is 10^{11}–10^{12} cm^{-2}eV^{-1} by Alexe [461]. However, this number increases rapidly when the device is annealed at high temperatures.)

Note that ferroelectric films are also directly deposited on Si substrates for micromachines [462]. However, for this application the interface trap density is relatively unimportant, compared with FET use [463].

13. Ferroelectrics-on-Superconductor Devices: Phased-Array Radar and 10–100 GHz Devices

The first invention using high-T_C superconductors on ferroelectric films (or high-dielectric paraelectric materials such as strontium titanate or BST that are nearly ferroelectric) were from the University of Colorado group involving Hermann, Yandrofski, Price, Barnes, and Scott [464, 465, 466]. These devices (Figs. 13.1and 13.2) incorporated the superconductor merely as a ground plane, with the ferroelectric film in transverse geometry (interdigitalized electrodes, no field across the film thickness) and a micro-stripline configuration (Figs. 13.3 and 13.4) [467]. The fundamental advantage of such devices was in their ability to provide very large phase shifts ($>$ 25%) at very low voltages, and to operate in the 10–20 GHz regime at low loss ($Q > 1000$, loss tangent $<$ 1%) [468]. Prior to this work, ferroelectrics had traditionally been viewed as unsuitable for microwave devices [469]. Hermann's group successfully fabricated devices on both YBCO (Yttrium Barium Copper Oxide) and the thallium-based high-T_C superconductors. A variety of devices were made and characterized by Galt, Price, and Ono [470].

Figure 13.6 shows the dependence of quality factor Q (reciprocal loss tangent) upon reduced temperature $\tau = (T_C - T)/T_C$ for BST having T_C near 100 K. Note that these devices are intended for 77 K or 4 K operation to optimize the use of the superconducting ground plane (make their sheet resistance negligible). Figure 13.7 shows the dependence of loss tangent upon film thickness. This is probably due to a damaged surface/interface layer. Kaiser et al. report direct evidence of such surface layers in BST (both pyrochlore and amorphous regions); and anatase is found in Bi-deficient or Pb-deficient titanates [471]. The intrinsic loss mechanism theory of Tagantsev and Gurevich predicts that the dielectric loss as a function of voltage scales as the capacitance $C(V)$ to the 3/2 power. This was confirmed for strontium titanate at 14 GHz and 77 K up to 5 V [470].

The inferred thickness of the damage layer from dielectric loss measurements is compatible with the estimate from the temperature dependence of the real part of the dielectric constant. For PZT (Fig. 13.8) the $\varepsilon(T)$ data satisfy the model of a damaged surface layer of 20 nm thickness and a temperature-independent dielectric constant of 200, in series with a bulk dielectric having Curie–Weiss behaviour and bulk, single crystal parameters. As in BST, this interfacial damage layer next to the electrodes is probably

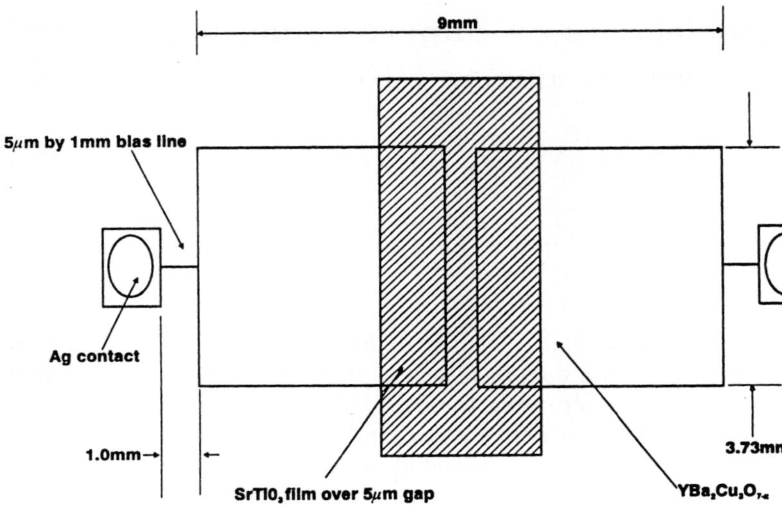

Fig. 13.1. BST-on-YBCO extended antenna for 14 GHz operation from Galt et al. [468]

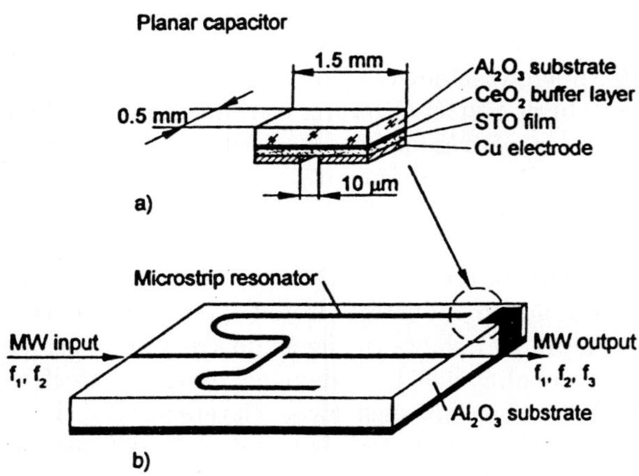

Fig. 13.2. BST-on-YBCO X-band device from Petrov et al. [486]

amorphous. In BST, Zafar et al. [470] showed that quantitative analysis of the frequency dependence $C(f, T)$ was possible by assuming a damaged surface layer of frequency-independent capacitance of about 1.0 nF in series with the undamaged interior part of the film.

Because this system produces very large phase shifts (Fig. 13.5) at very low voltages, it can be used as the phase-shifter in phased-array radar [470, 472]. For readers unfamiliar with these devices, they are stationary antennas (unlike the rotating dishes) which direct the radar beam by phasing the

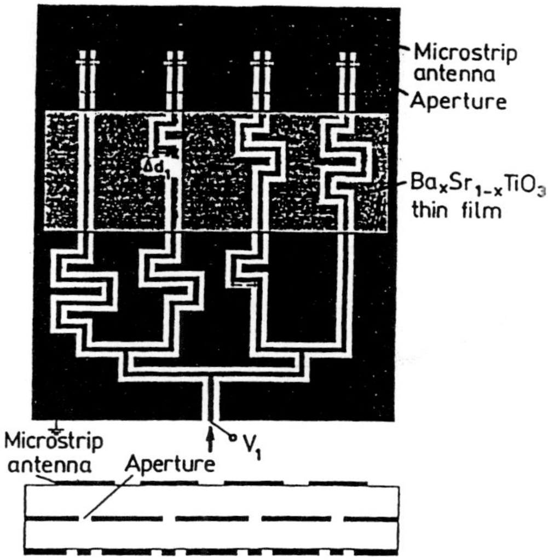

Fig. 13.3. Electrode feed system for phased-array radar with $SrTiO_3$ or BST-on-YBCO, Barnes et al. [467]

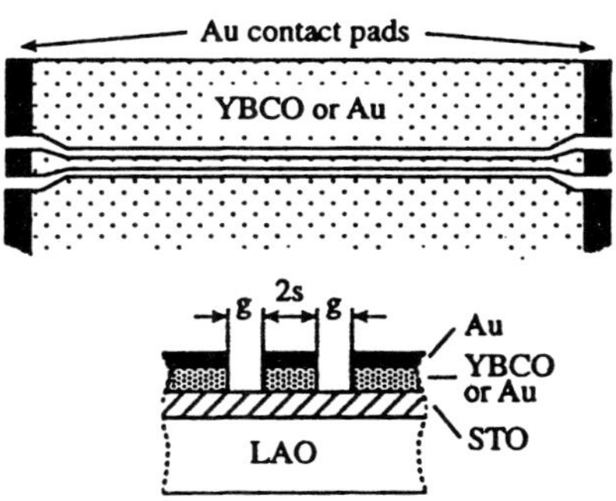

Strip width (2s) / Gap width (g), [μm]			
4/18	6/25	14/41	30/75

Fig. 13.4. Phased-array device with $SrTiO_3$ or BST-on-YBCO, Kozyrev et al. [487]

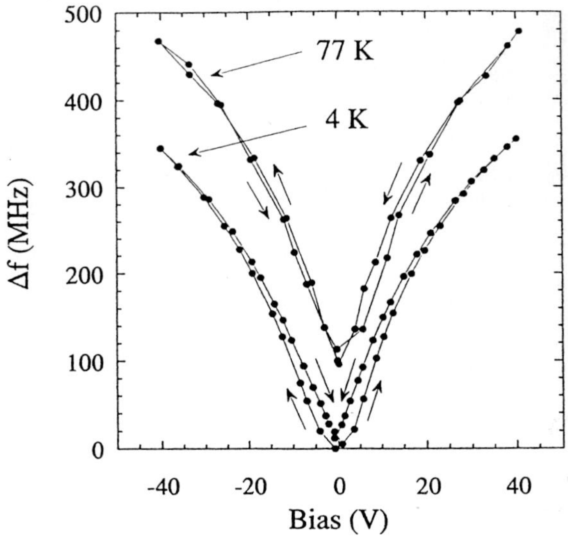

Fig. 13.5. Phase shift measured as frequency, versus applied voltage V at 4 K and 77 K in the device shown in Fig. 13.1 [468]

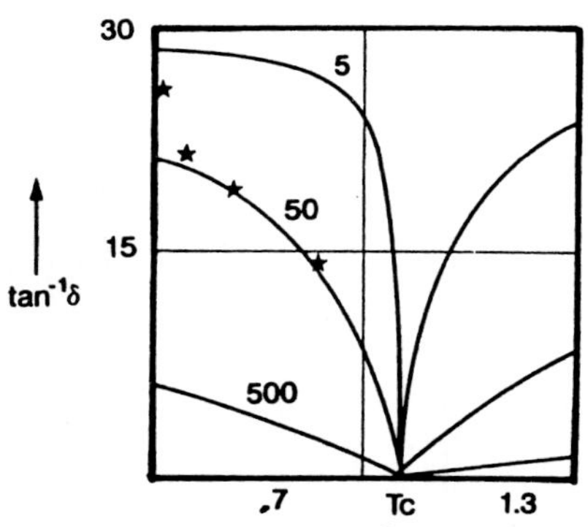

Fig. 13.6. Reciprocal loss tangent (quality factor Q) versus reduced temperature $\tau = (T_C - T)/T_C$ in BST films at 14 GHz [470]

applied voltage to produce a sweep of nearly 180 degrees of arc. Typically 50° at 2.4 V/μm at 4 GHz with a small insertion loss of 4 dB is obtainable. TRW reports 80° at 10 GHz at 30 V and 140° at 40 V over a PLD SBT film and only 3 dB insertion loss at 20 GHz. The advantage of ferroelectric thin-film devices for this application is that existing phased-array radar devices are relatively heavy, bulky, or require high voltages. As with memory devices, integrated ferroelectric thin-film systems, will be lighter, smaller, rather robust, and work at very low voltages (a few volts). One volume of a journal has recently been devoted to this particular application of ferroelectric thin films [473],

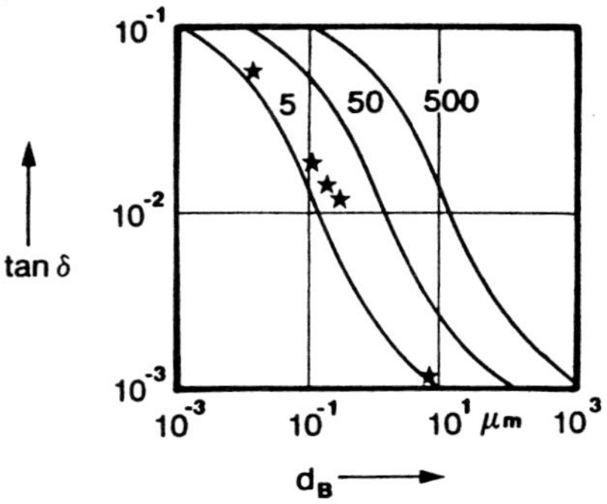

Fig. 13.7. Loss tangent at 14 GHz versus film thickness d in BST films [470]

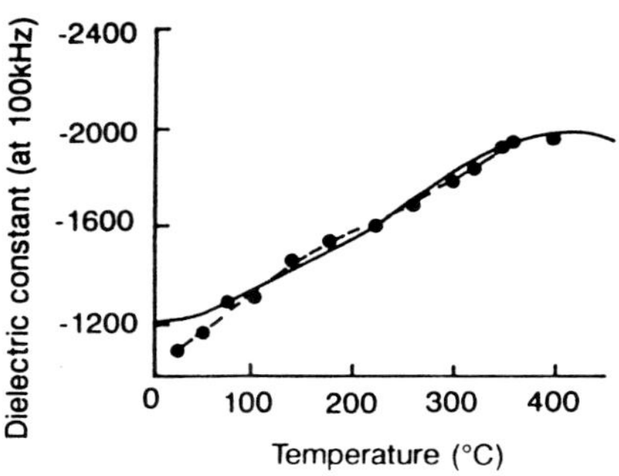

Fig. 13.8. Dielectric constant versus temperature in PZT. The *solid line* is a fit to a theory [470] with a surface layer of thickness 20 nm and temperature-independent dielectric constant of 200, on top of a film characterized by (1.5) with bulk properties: $\varepsilon_{\mathrm{b}}(T) = 10^6/(T_{\mathrm{C}} - T)$, $T_{\mathrm{C}} = 670$ K and $\varepsilon\,(293\text{ K}) = 1300$

but since this book emphasizes memory applications, further discussion is not provided here. Readers are referred to the patent literature [474, 475]. Rather good work has been done at Grumman [476], TRW [477, 478], and at US Army laboratories [479, 480, 481, 482] for the 35 GHz regime.

In the future we might expect a wider variety of ferroelectrics to be used in such low-temperature applications. Combined with superconducting ground planes operated at 77 K, many materials that are not ferroelectric at room temperature could be used and therefore should be explored. Even near 4 K there are materials which have rather high dielectric constants because their Curie temperatures are only slightly below 4 K (such as pure strontium titanate, which could be used as a 4 K liquid-helium DRAM capacitor of the future and provide $\varepsilon = 30\,000$ or greater); and there are ferroelectrics whose Curie temperatures are only slightly above 4 K, albeit with large activation switching fields. This conflicts strongly with the view of Tokunaga [483, 484] that there are no crystals with Curie temperatures below 100 K. A list of

Table 13.1. Low Temperature Ferroelectrics

Formula	Curie temperature T_C	Curie constant C
$NH_4Al(SO_4)_2 \cdot 12H_2O$	71 K	?
$NH_4Fe(SO_4)_2 \cdot 12H_2O$	88 K	400 K
$(NH_4)_2Cd(SO_4)_3$	95 K	?
$CdTiO_3$	55 K	4.5×10^4 K
$Pb_2Nb_2O_7$	15.3 K	?
$LiTlC_4H_4O_6 \cdot H_2O$	10.5 K	?
$K_3Li_2Nb_5O_{15}$	7 K	?

such materials [485] is given in Table 13.1 in the hope that it will stimulate further work on such liquid-helium cooled devices.

14. Wafer Bonding

Wafer bonding is a technique whereby a film, including a ferroelectric film, is grown on a sacrificial substrate which is subsequently removed by etching it off from the back [488]. It avoids the exposure of the ferroelectric semiconductor to high processing temperatures [489], and hence is especially applicable to systems such as ferroelectric-gated FETs (Chap. 12) or pyroelectric detectors using PST (lead scandium tantalate) – which suffers from a very high (900°C) processing temperature that prevents full integration into Si chips. (However, recently PST has been made via sol–gel at lower temperatures [489].)

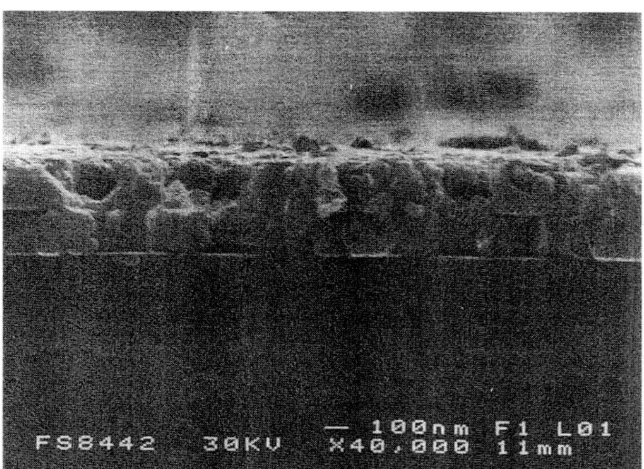

Fig. 14.1. Cross-sectional SEM of PZT/Si after the sacrificial bottom Si wafer was etched away [489]

When PZT thin-film memory devices are made via wafer bonding, they are usually prepared by chemical solution deposition and subjected to rapid thermal anneal (1000 K for 60 s). Then they are polished by chemical and mechanical techniques and bonded directly onto silicon wafers at room temperature. These bonded wafers are then annealed for about 12 hr at relatively low temperatures (473 K to 723 K). The bonding energy increases from an initially very small value of 100 mJ/cm^2 before annealing to around 1.4 J/cm^2 at the highest annealing temperature. Direct metal–ferroelectric–silicon films (suitable for FE-FETs) were then obtained by thinning down and etching away the Pt-coated silicon wafer. The films shown in Fig. 14.1 were made with a standard methoxyethanol route [490] using lead acetate. Deposited on 3″ Pt/Ti/SiO2/Si wafers, via spin-on at 2000 rpm, these were rather conventional PZT/Pt films. The subsequent wafer-bonding process proceeded as follows: chemical–mechanical polishing with Syton and glycerol on Chemcloth and a Logitech polishing machine, and cleaning for 1 hr in deionized water (to remove Syton residue). The PZT/Pt/Si wafer was then bonded under hydrophilic conditions to a second (top) Si wafer by standard wafer-bonding steps [491]. These steps consist of placing the top wafer on the deposited film in a highly parallel stack via three small spacers, and then blowing any dust out from between the two wafers with a spray of deionized water. After drying the water by spinning (2500 rpm for 5 min) under an infrared lamp at 373 K, the spacers are removed and the clean Si wafer is dropped onto the bottom PZT-coated one. Bonding occurs via simple finger pressure on the middle of the outside surfaces of the two Si wafers.

Bonding can be verified via an infrared video camera and light source; and the bonding energy can be measured via prying them apart [492].

After this bonding, the original Si wafer is polished down to 50 microns and then etched in KOH, resulting in the direct PZT/Si structure shown in Fig. 14.1. The Pt/SiO_2 layers act as an etch stop. The latter layers can explosively blow off the PZT, due to the fact that their bonding energy is less than that of the PZT to top Si surface.

The final PZT/Si (top wafer) structure is then metallized with aluminium on the back side of the Si wafer after the SiO_2 layer has been removed with HF, and Au on the top (patterned with a mask set).

The resulting device is a rather good PZT FET, with a memory window of 2.8 V and good accumulation and inversion voltage behaviour. Breakdown is at 30 V across 450 nm (Fig. 14.2).

The advantage of applying these techniques for PST detectors is that they have a pyroelectric detector figure of merit five times as great as that of PZT [493]. But it is processed at 1173 K, precluding full Si integration (only hybrid solder-bump structures). The resulting structure is not competitive with cantilevered PZT/Si detectors [494], although very recent results by Takeishi and Whatmore [489] suggest that this high-temperature processing limitation can now be overcome.

It is my view that wafer bonding will be a very useful tool for making FE-FETs and for pyroelectric detectors. Perhaps one problem to delve into further is the risk of hydrogen contamination with water-based processing

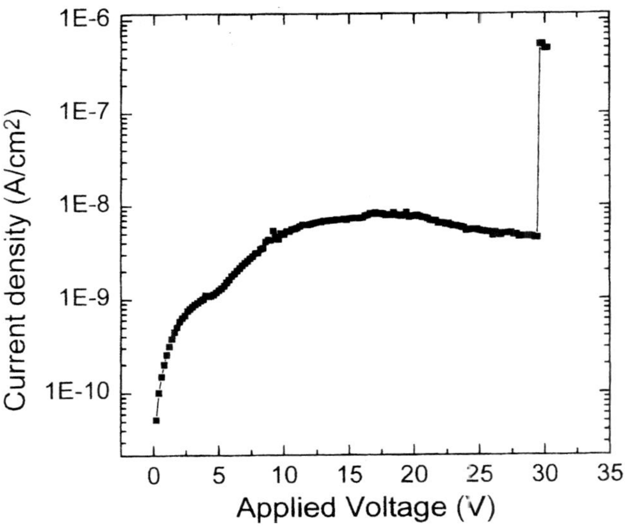

Fig. 14.2. Electrical breakdown behaviour of Au/PZT/Si (200 s dwell time). At $d = 450$ nm, the breakdown field E_B is 0.67 MV/cm [489]

(cleaning, etc.); but alternative water-free techniques are already available. The other problem is that it is not yet certain what the trap concentration will be at the ferroelectric–silicon interface. Unless this is small, commercial FE-FETs cannot be fabricated in this way.

15. Electron-Emission and Flat-Panel Displays

When ferroelectrics are subjected to applied electric fields sufficiently large to switch their domains, electrons are emitted from their surfaces [495, 496]. This process is related to the Barkhausen effect, in which the annihilation of two colliding charged 180° domains liberates free charge [497]. Although regarded as a novelty for academic laboratories for many years, more recent work by Gundel et al. [498, 499] has shown that this emission can be synchronized to very fast rise-time electric pulses (ns) and therefore provide a useful pulsed electron source for accelerators. Because of the very short pulse widths, Gundel has achieved very large current densities (of order 1 A/cm^2), as shown in Figs. 15.1 and 15.2, using a rather simple structure (Fig. 15.3) with 200 micron grids.

At this stage this phenomenon is under study [500, 501, 502, 503] for flat-panel display devices, where liquid crystal displays work extremely well but have low yields in very large sizes. The group at Nantes has developed microchannel-plate prototypes. The main problem at present is that the ferroelectrics used (e.g., PZT) work much better in bulk than in thin-film form. The most recent developments reported are by Benedek et al. [504].

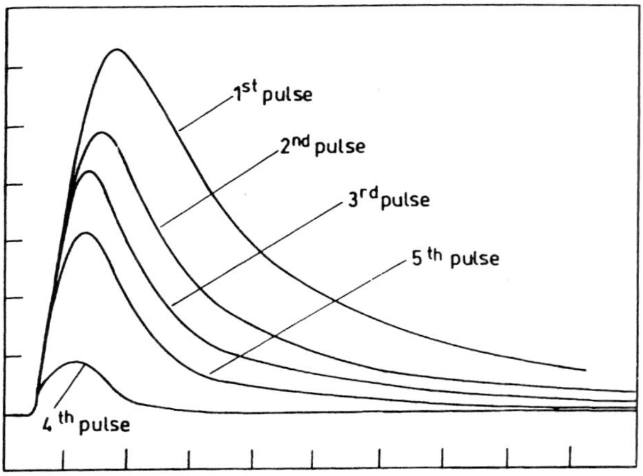

Fig. 15.1. Nanosecond emission charge densities from the surface of PZT versus time t [503]

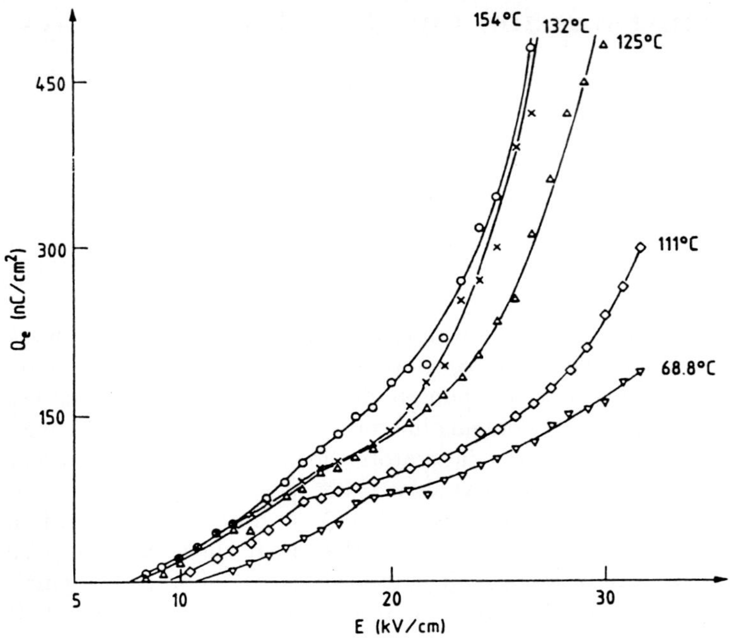

Fig. 15.2. Surface emission from PZT versus temperature T [503]

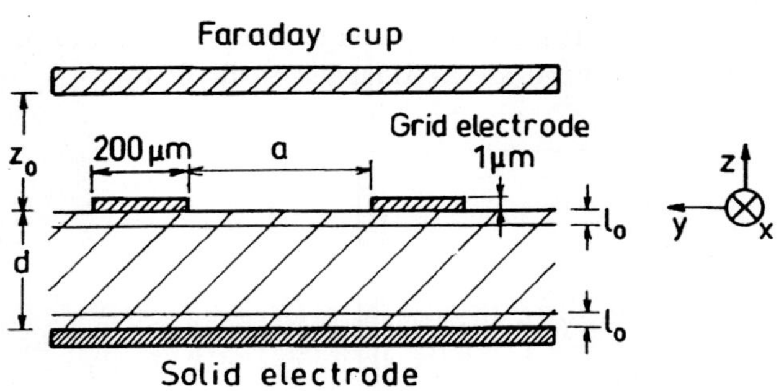

Fig. 15.3. Experimental configuration for electron emission studies from the surfaces of ferroelectrics [503]

16. Optical Devices

Optical devices are an important part of this book for they are not peripheral to the basic development of ferroelectric memories. Not all memories will have electrical READ and electrical WRITE operations. Several researchers have already investigated electrical READ/optical WRITE FE-RAMs with the intention of using nondestructive optical READs to eliminate reset and hence greatly reduce fatigue [505, 506, 507, 508]. Optical WRITE/electrical READ devices are potential replacements for X-ray film. At present, X-ray records take a lot of hospital storage space (or are kept in patients' homes) and are time consuming and expensive to retrieve and deliver to the doctor's office in a timely manner. Moreover, reading X-ray films and comparing them is an art. It would be a great improvement to have these films available digitally on the PC (Personal Computer) of every doctor's office, in high resolution (> 4 megapixel) form, with grey scales (obtained from different bias voltages), and the ability to align different X-rays and digitally subtract one from another, to show only changes in the images (e.g., tumor size). Finally, all-optical READ/WRITE memories (*photonic* devices) may become possible with ferroelectrics. These possibilities are discussed below.

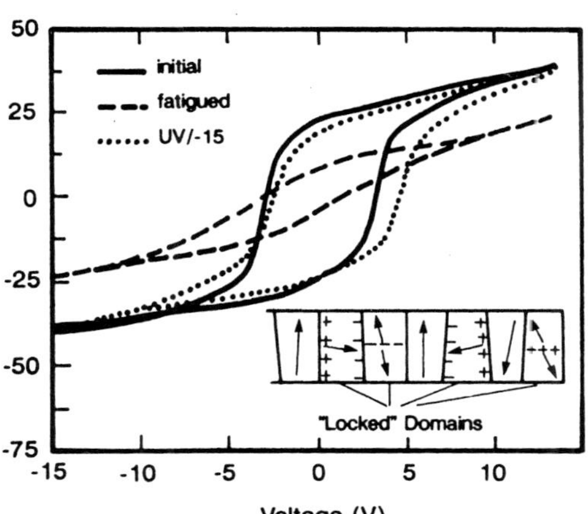

Fig. 16.1. Shift in coercive voltage for ferroelectric PZT films under optical illumination [507]

A very simple optical ferroelectric of great potential importance is the birefringent optical bistability discovered by Haertling's group at Clemson [509, 510, 511].

A different effect is the shift in coercive voltage or field in the presence of near-bandgap optical illumination in PZT studied by Warren et al. [512] and illustrated in Fig. 16.1.

It is useful to keep in mind that the original ferroelectric memory devices of Wu [513] could be read optically (via beam deflection).

The use of bismuth titanate thin films for optical (UV) detectors has been proposed recently [514]. And it has been pointed out that the use of a sandwich (bimorph) of two ferroelectric films together can make a superior infrared pyroelectric detector [515].

17. Nanophase Devices

17.1 Lithography

The best nanoscale ferroelectric thin-film arrays to be produced thus far are by Alexe [516]. As shown in Fig. 17.1, these e-beam arrays of SBT are 100 nm on a side and switch very well. They are about 100 times smaller than those reported elsewhere [517]. These arrays were fabricated with minimum lateral sizes less than 100 nm by using Electron Beam Direct Writing (EBDW); such maskless direct writing eliminates the need for submicron etching. It is useful for metallic and oxide nanostructures. Chemical reactions are induced locally in a metal–organic thin film by irradiation with an e-beam. The resulting pattern is developed by dissolving the unexposed area in a specific solvent and fixed by thermal annealing.

In the data illustrated in the accompanying figures, a commercial e-beam lithography system (ELPHY Plus) was adapted to a JEOL JSM-6400 SEM working at 40 keV. After developing, the metal–organic mesas were transformed to oxides by annealing in air for 5 min at 573 K, and then cystallized into the proper ferroelectric phase by a second anneal at 873 K to 1123 K. The resulting cells are sufficiently small to serve in 1 Gbit arrays. During the crystallization process the SBT cells lose their square shape, but PZT retains it even for 50% shrinkage.

The ferroelectric switching for similar self-assembled bismuth titanate was shown earlier [518, 519]. The arrays themselves (SEM micrographs) are shown in Figs. 17.3 and 17.4. The leakage current for these bismuth titanate devices is space charge limited, as shown in Fig. 17.5. E-beam lithography to micropattern 1 micron arrays of bismuth titanate has also been reported by Okamura et al. [517], but Alexe's are 100 times smaller in area. However, after sintering in oxygen for 30 min at 1073 K, Okamura achieved feature sizes around 0.2 microns and 0.3×0.3 micron square cells. Okamura's work is also interesting in that it utilizes bismuth octalate as the Bi precursor.

17.2 Surface Droplet Epitaxy

The production of the high-temperature, conducting, δ-phase of bismuth oxide on bismuth titanate and SBT films was a surprise [518, 519]. The fact

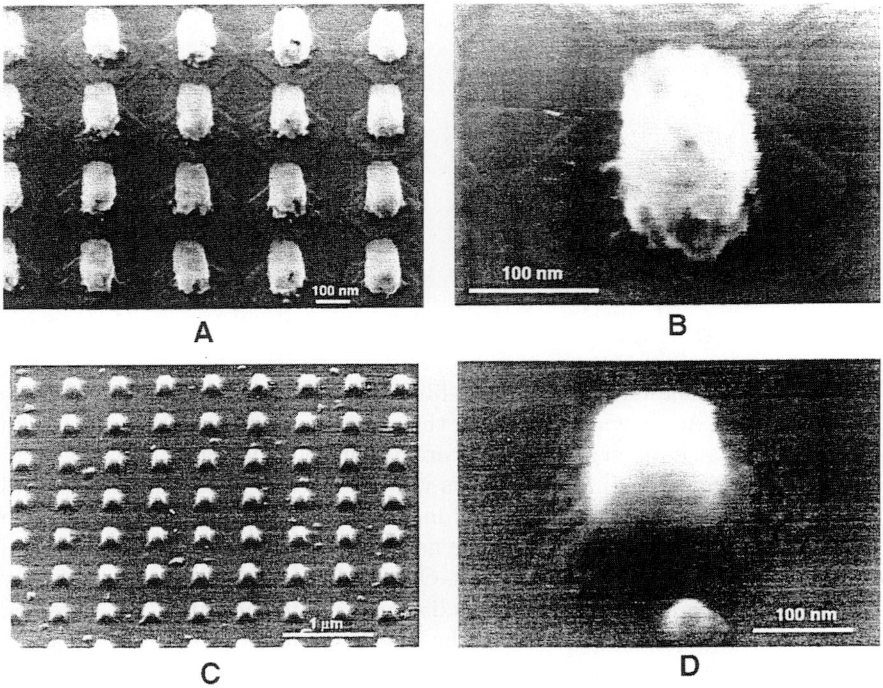

Fig. 17.1. 100 nm SBT ferroelectric cells [516]: (**a**) just after developing (unheated); (**b**) smallest features in (**a**); (**c**) ×45 000 magnification after 850°C crystallization; (**d**) ×95 000 magnification after crystallization

that thin films of the δ-phase of Bi_2O_3 are stable at room temperature, discovered by Alexe et al. in 1997, was independently rediscovered by Switzer et al. in 1999 [519]. Bi_2O_3 nanomaterial as small as 10 nm in diameter has been prepared by Yu et al. [520]. For these very small samples there is a large confinement energy of 0.2 eV. But since the confinement energy varies as d^{-2}, in our 100 nm specimens, this is a negligible 1 meV or so.

We note that the size of the bismuth liquid droplets that nucleate on the top surface of SBT films during their high-temperature processing and subsequently oxidize to form δ-phase Bi_2O_3 can be estimated from the Raleigh–Thomson instability criterion [521] (17.1). This expresses the minimum size r_{min} that a droplet can have if subjected to a voltage V. We take V here to be the contact potential 0.8 V between the SBT electron affinity $\chi = 3.4$ eV, and the Bi metal work function, $\Phi_{\mathrm{M}} = 4.2$ eV:

$$V < 4(\pi s r_{\mathrm{min}})^{1/2} , \tag{17.1}$$

where s is the known surface energy [521] for liquid Bi of 53 erg/cm^2. Relation (17.1) yields r_{min} of about 0.3 microns, in reasonable agreement with the observed value of 0.1 μm. The idea that this self-assembling process involves

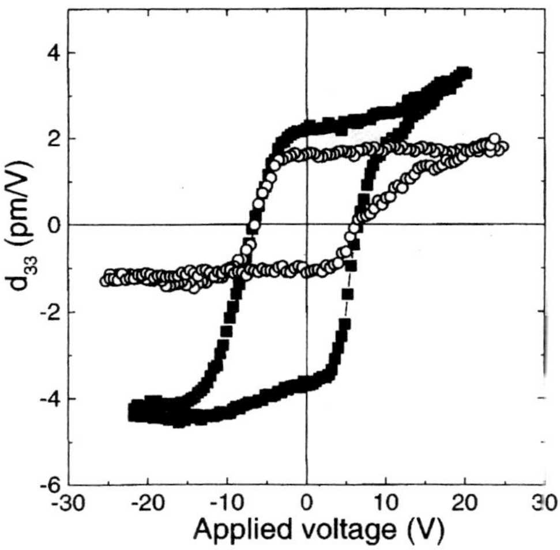

Fig. 17.2. Hysteresis for 100 nm diameter self-assembled SBT cells (*open circles*) [519]. 100 × 100 nm PZT (*solid squares*). Note that compared to the 1 micron cells denoted by *solid squares*, the 400 × 250 nm cells have the same coercive field E_c and the same near-zero bias voltage V_{bi}, but they have a different vertical shift – positive instead of negative. This arises from space charge accumulation at the top or bottom electrodes (which are different). (Figure from A. Pignolet and M. Alexe)

Fig. 17.3. Structure of bismuth titanate nanoscale δ-Bi_2O_3 electrodes on bismuth titanate [519]: plan view. Si axes refer to the underlying silicon single-crystal wafer

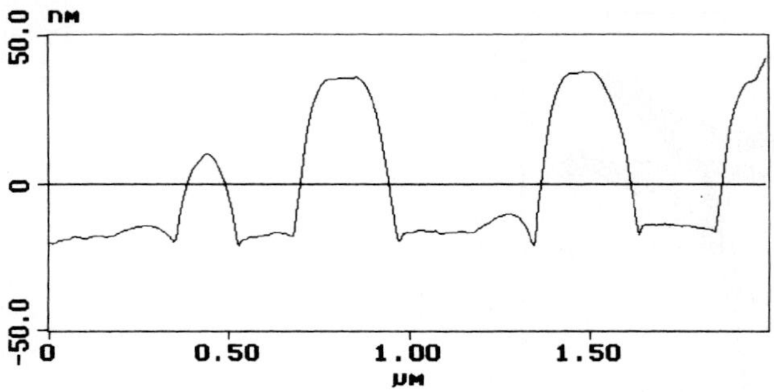

Fig. 17.4. Structure of bismuth titanate nanoscale δ-Bi_2O_3 electrodes on bismuth titanate [519]: cross-sectional view

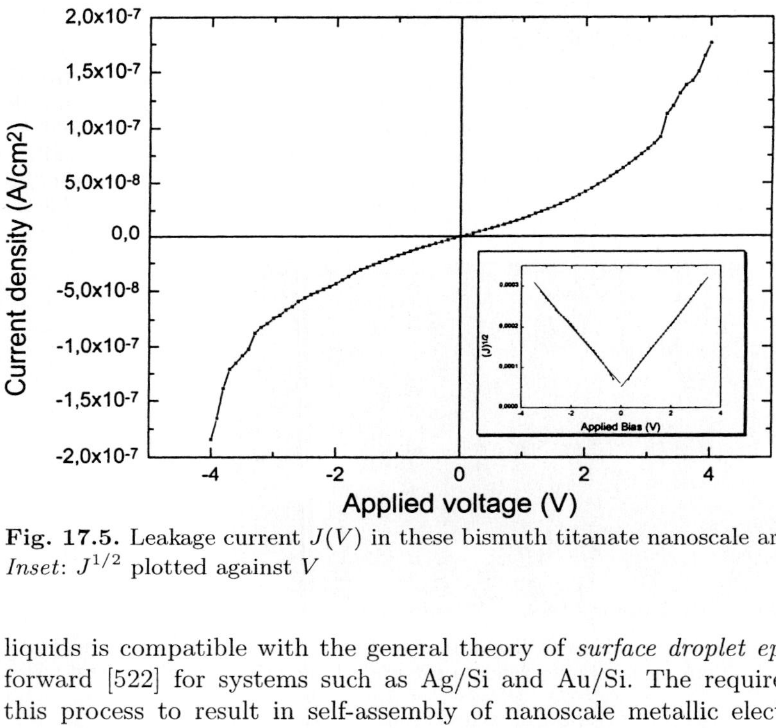

Fig. 17.5. Leakage current $J(V)$ in these bismuth titanate nanoscale arrays [518]. *Inset*: $J^{1/2}$ plotted against V

liquids is compatible with the general theory of *surface droplet epitaxy* put forward [522] for systems such as Ag/Si and Au/Si. The requirement for this process to result in self-assembly of nanoscale metallic electrodes on semiconductors is that there exist an exothermic eutectic phase transition within the processing temperatures used.

Epitaxial films of all the Aurivillius-phase materials can be made. A scanning force micrograph of bismuth titanate is shown in Fig. 17.6. TEM cross sections (Fig. 17.7) reveal that these films actually consist of two sublayers: the lower sublayer (right on the electrode) consists entirely of c-axis grains;

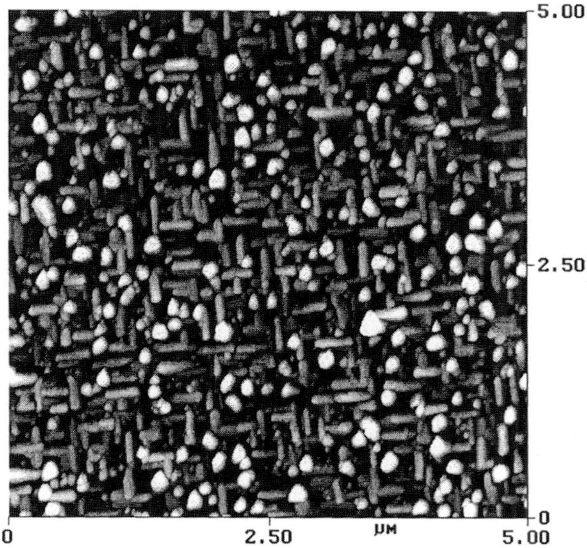

Fig. 17.6. AFM topographic image of rectangular (110) a_t-grains protruding out of a flat (001) c-surface in barium bismuth titanate. Depth is from zero (*black*) to 75 nm (*white*). Lateral scale in microns

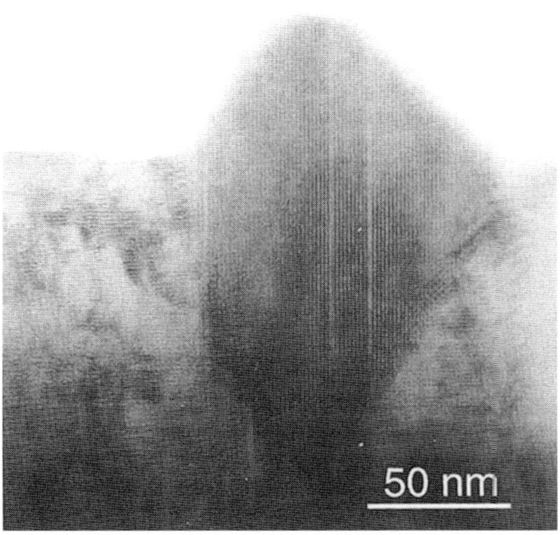

Fig. 17.7. TEM cross section of a (110)-grain in SBT protruding out of a flat (001) c-surface, similar to that in bismuth titanate in Fig. 17.6

but the upper layer has mostly a_t-axis (110) grains embedded in a c-axis matrix.

Ideally the self-patterned metal electrodes for ferroelectric memories would be a metal already in the ferroelectric, excreted during processing to form a registered array of nanoelectrodes. The discussion above shows that Bi from SBT works. However, it would be equally good if Ti could be similarly self-patterning on PZT, and this does not work. Ironically, Ti is known to have the right combination of physical properties to self-pattern [523]. Registration techniques for Ti nanodots have been described [523]. Perhaps additional work on this would be merited. At present the Bi nanotechnology is viewed as a radical departure from studies of Ti or Ag self-patterning in microelectronics and as a controllable phenomenon in which switching of submicron ferroelectrics has already been achieved. These schemes are of course unrelated to colloidal monolayer techniques [523].

17.3 Use of Bi Surface Droplets to Optimize SBT Stoichiometry

Rapid thermal annealing initiates the crystallisation of bismuth layer structured ferroelectric thin films. Subsequent crystallisation annealing for longer periods improves the crystallinity of the thin films. During annealing bismuth is known to diffuse to the surface and into the electrode which is deleterious to the device performance. Forming gas annealing subsequent to the rapid thermal annealing (prior to the crystallisation annealing) significantly reduces bismuth diffusion into the electrode. The suppression of bismuth diffusion into the electrode films and the elimination of excess bismuth in the films improves and stabilises the electrical properties of the capacitors, in particular the electronic conduction.

As discussed earlier, bismuth layer-structured ferroelectric thin films have become such promising candidates for use in nonvolatile random access memories due to their lack of fatigue character and low coercive field. $SrBi_2(Ta_xNb_{1-x})_2O_9$ (SBTN) thin films particularly show a great performance. However, excess Bi fluxes are required to ensure the proper phase formation and the good electrical properties in SBTN thin-film growth. This excess Bi remains in the films even after SBTN formation and has a great influence on their electrical properties. The annealing temperature ($> 700°C$) for the SBTN crystallisation is high enough for the diffusion of Bi into Pt electrodes and the formation of Bi–Pt alloys. This change in interface affects the ferroelectric–metal contacts. To make the film performance stable, especially the current–voltage dependence, the ferroelectric–metal contacts should be stable.

Several attempts to remove excess Bi have been performed using a Rapid Thermal Annealing (RTA) process. Watanabe et al. (Sony) showed [524] that the SBTN phase could be formed completely by only RTA, but additional heat treatment was required to obtain good electrical properties.

Recently Zafar et al. (Motorola) found [525] that the forming gas annealing caused the formation of Bi columnar structures on the film surface and they were removed after an oxygen anneal at 800°C.

Watanabe et al. [526] describe a method to suppress the Bi diffusion in the SBTN thin-film growth and to provide stable ferroelectric–metal contacts.

Oxidation of silicon wafers

Etch the back sides of Si wafers to remove SiO_2

Deposit bottom electrode (200 nm Pt/ 30 nm Ti) onto SiO_2/Si

Spin coating of Bi-rich SBT (Sr/Bi/Ta = 0.7/2.4/2/0)

Drying (525 K for 400 s)

Rapid thermal anneal (30 s in O_2)

[Repeat preceding three steps until desired thickness is attained.]

Crystallization anneal in O_2 at 1073 K for 1 hour

Top electrode deposition

Postanneal

Fig. 17.8. SONY process for SBT thin films

Figure 17.7 shows the SONY process for the $SrBi_2Ta_2O_9$ (SBT) thin film. The solutions were spun onto Si/SiO_2/Ti(30 nm)/Pt(200 nm) substrates. After spin-coating, the films were dried at 250°C and rapid-thermally annealed in oxygen atmosphere for 30 s. After this procedure was repeated three times in order to obtain the required film thickness, the films were heated in oxygen atmosphere at 800°C for 1 hr to promote the crystallisation of the SBT. The final film thickness was 180 nm. Figure 17.9 shows the result of XRD measurements after RTA process (before crystallisation annealing at 800°C). RTA temperature was varied from 690°C to 785°C. At lower RTA temperature, a fluorite structure was formed, which was the expected intermediate structure of the desired Aurivillius phase of SBT. However, the Aurivillius SBT structure became dominant at higher RTA temperatures. Above 760°C, the films be came completely Aurivillius layer-structure single phase. RTA here consisted of a heat treatment for a short period with high ramp rate (usually more than 100°C/s). The RTA processing is thought to minimise the diffusion of the film elements into the substrates, compared to furnace

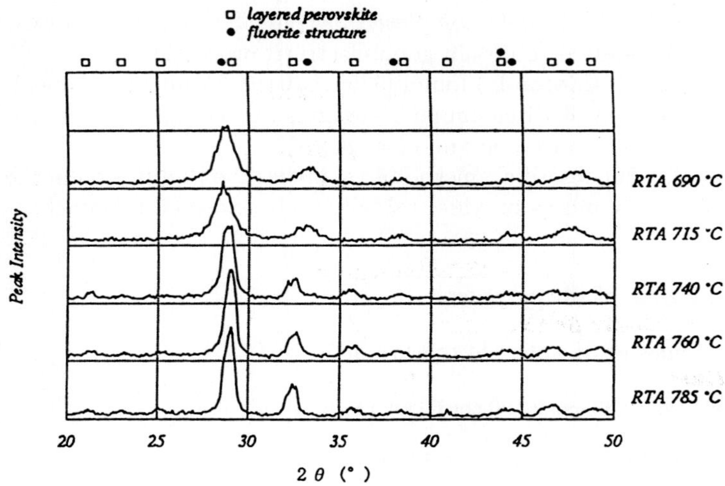

Fig. 17.9. XRD X-ray diffraction results on spin-on SBTN samples

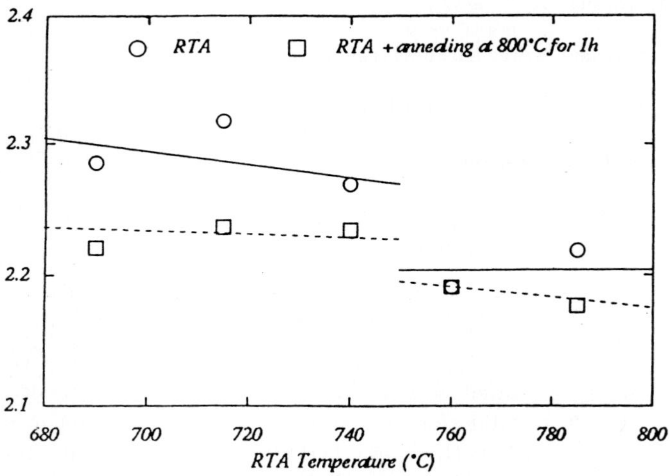

Fig. 17.10. Discontinuity in Bi percentage in SBT films versus processing temperature, presumably due to BiPt formation at the electrode interface

annealing, because the high ramp rate causes rapid crystallisation of the films. Therefore in our SBT films, Bi diffusion into the Pt electrodes is thought to be minimised after the SBT formation (above 760°C). Note in Fig. 17.10 that Bi reacts with Pt at around 740°C to form BiPt. Recently Siemens/Symetrix have reported extra XRD lines in SBT films after patterning and conclude that etching causes a reaction of SBT with Pt at the sidewall. Annealing

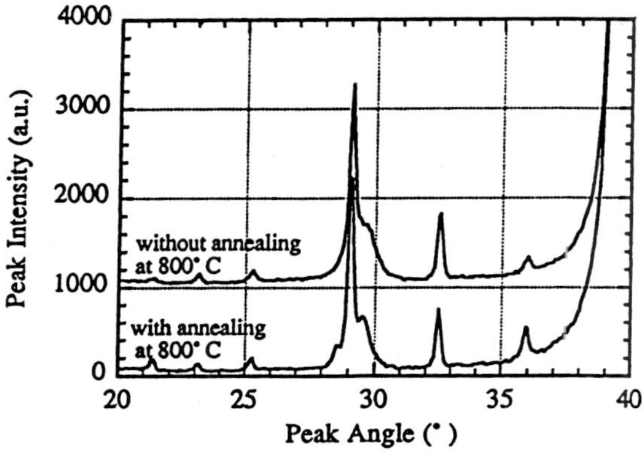

Fig. 17.11. XRD of SBT before and after 1 hr anneal at 800°C

the top Pt electrode before patterning improves the Pt crystallization and minimizes this problem [527].

Additional heat treatment at 800°C for 1 hr in oxygen flow was required to ensure good electrical properties. Figure 17.11 shows the comparison of XRD patterns between the SBT films without annealing at 800°C for 1 hr and those with annealing at 800°C for 1 hr. Both films were rapid-thermally annealed at 760°C. A Θ–2Θ scan mode was used in these measurements. It can be seen that the SBT films with annealing at 800°C have sharper diffraction peaks, which means better crystallinity. Conversely, the SBT films after RTA process are thought to be at some midpoint in their growth. Put another way, the crystal growth does not saturate. It also becomes obvious from SEM images.

Figure 17.12 shows our new process for SBT thin-film growth. A forming gas 'recovery' annealing is introduced after RTA process. A forming gas annealing here means a heat treatment at 400–500°C for 0.5–1 hr in 5% H_2 and 95% N_2 flow to recover the transistor performance. The species of gases are not limited to forming gas at these percentages. Other reductive ambient gases may be used in this process.

The most important thing here is to introduce a forming gas annealing before the crystal growth of the SBT saturates. According to Zafar's report, a forming gas annealing at 500°C for 1 hr caused Bi columnar formation on the SBT surface and a 16% decrease in film thickness (see Fig. 17.13). This columnar structure could be removed by an additional annealing at 800°C for 1 hr in oxygen flow because this oxygen anneal caused Bi to form Bi oxide which vaporized at this annealing temperature. However, they reported that craters were left in place of the Bi islands and that the regions under the craters were amorphous because of the lack of Bi (see Fig. 17.13). These problems can be resolved in our new process. In our process, a forming gas annealing is in-

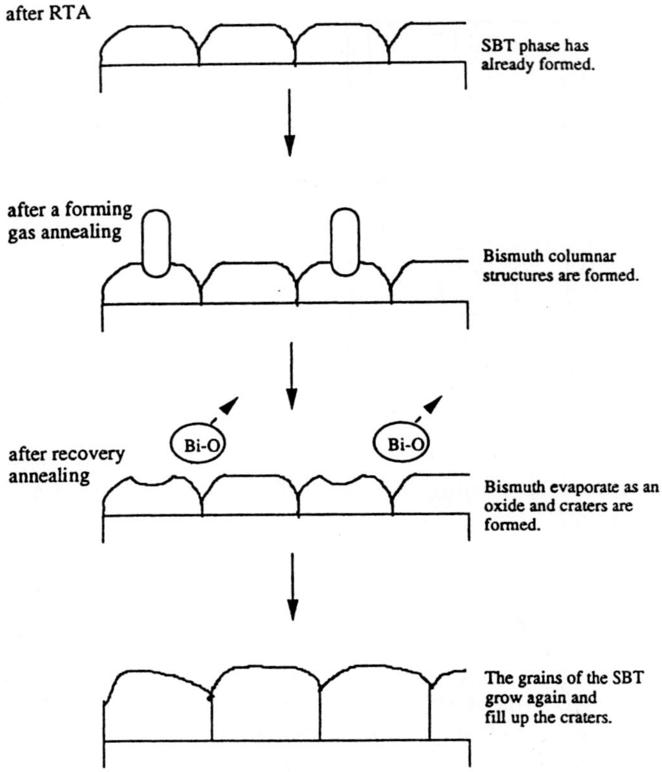

Fig. 17.12. The new process to optimize SBTN stoichiometry of bismuth [526]

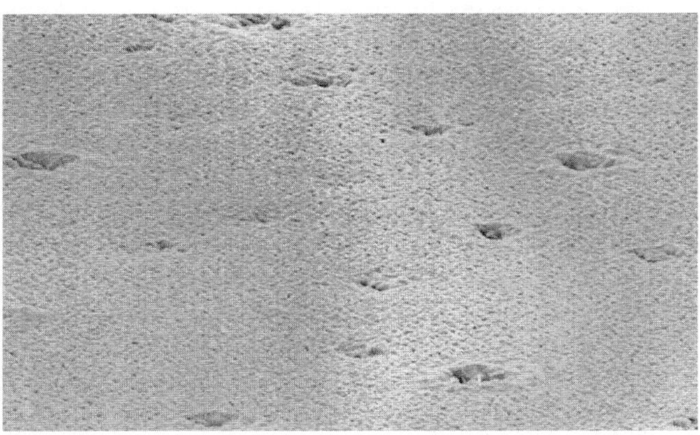

Fig. 17.13. Amorphous region under craters where Bi islands were removed in the SONY process of Fig. 17.8: plan view from Motorola [525]

troduced before the crystal growth of the SBT saturates. Therefore the films continue to grow during a recovery annealing at 800°C in oxygen flow after a forming gas annealing. The formation of craters can be avoided here. Moreover, the amount of Bi removed during a recovery annealing can be controlled by the parameters of a forming gas annealing (temperature, annealing time or annealing ambient). Amorphous regions originated from Bi deficiencies can also be avoided here.

This process can be employed with techniques other than spin-on. We mentioned the thin films prepared by a spin-coating method in the previous section; however, we can obtain the same effects from the thin films prepared by thin-film growth techniques other than a spin-coating method, such as a Chemical Vapour Deposition (CVD) or sputtering.

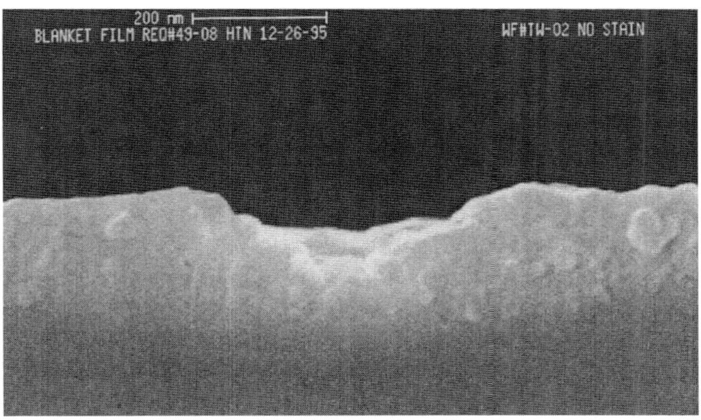

Fig. 17.14. Results of forming gas annealing on Bi diffusion in SBT Cross-sectional view [525]

Isobe et al. reported [528] that Aurivillius layer-structure single phase was obtained from an as-deposited amorphous film, which was deposited below 450°C using the flash-CVD system and then annealed at 800°C in oxygen ambient for 1 hr. Aurivillius layer-structure single phase can also be obtained by applying a RTA process on the amorphous thin films prepared by the methods mentioned above. Forming gas annealing prior to the crystallisation annealing (see Fig. 17.14) also reduces the Bi diffusion into the electrode, as in the case of the spin-coated thin films.

17.4 Fringing Fields

As both DRAMs and nonvolatile FeRAMs are reduced to submicron or even nanoscale lateral cell sizes in order to serve in Gbit memories, questions arise as to the ultimate minimum size possible for functioning cells. Recently Alexe et al. at the Max Planck Institute for Microstructural Physics in Halle have demonstrated that 0.1×0.1 μm^2 devices will switch [529]. So we know that there is no fundamental physics limitation on going to this nanoscale. However, it will be useful for designers and device engineers to understand the constraints at such small sizes.

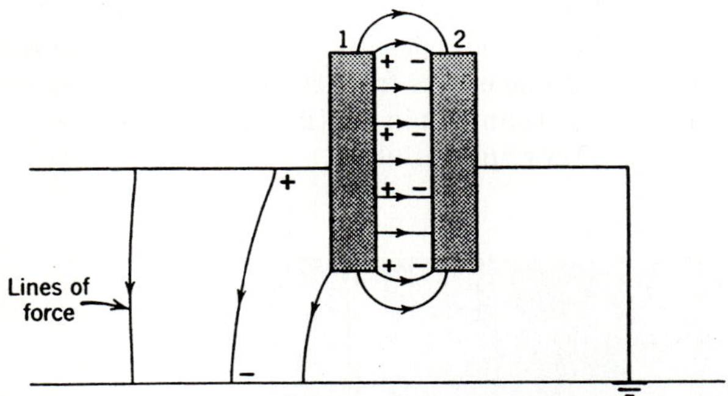

Fig. 17.15. Fringing fields in a parallel plate capacitor

The fundamental limitation is probably fringing fields. As shown in Fig. 17.15, fringing fields in a parallel plate capacitor are those lines of force or field that extend outside the region of space directly between anode and cathode and thereby fail to penetrate the dielectric, penetrating instead the sidewalls, which may be metallic, vacuum, or some other insulating material. Students sometimes think that these fringing fields are 'extra' and add to the total capacitance; in fact, the opposite is true. Fringing fields always subtract from the capacitance that would be achieved in their absence. This may be argued qualitatively in the following way. The lines of force outside the dielectric pass through a region of smaller permittivity, but the total charge q on the plate is constant. Therefore the capacitor behaves as if its charge were shared between two capacitors in parallel, one with a low dielectric constant. This reduces the total value of C. Even if the capacitor had a vacuum gap, its capacitance would be reduced by fringing fields; this is a different effect which can be seen intuitively from Fig. 17.15. The fringing lines of force travel a greater distance from anode to cathode. This is equivalent to having the capacitor edges farther away from each other than the real separation d of the plates.

Experimentally it is difficult to separate out true fringing field effects from materials processing problems that cause the film quality to be worse at cell edges than in the interior. However, where the latter has been monitored carefully, it would appear that real fringing fields are important. The best example is perhaps from NEC, shown in Fig. 17.16. Here the switched charge at 2 V applied is plotted versus voltage, and a logarithmic dependence on

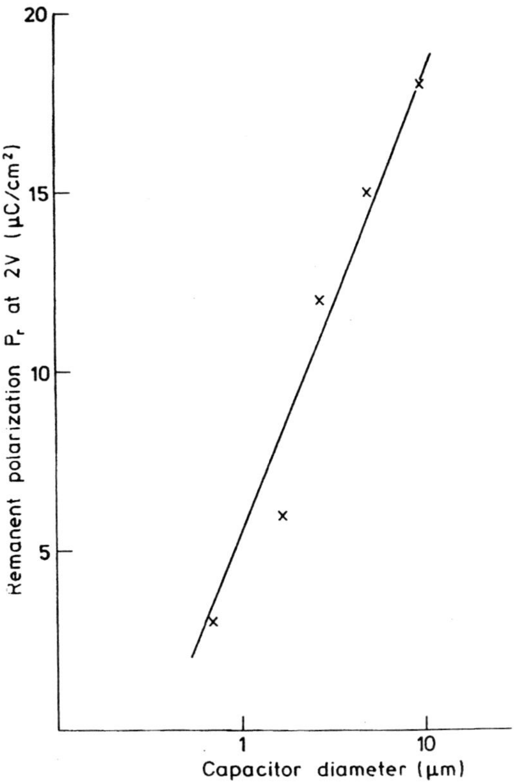

Fig. 17.16. NEC data on switched charge versus cell diameter for SBT arrays [529]

cell size A is observed [530]. For large rectangular cells the amount of charge switched q at a constant voltage V will be proportional to the area $A = ab$ of the rectangular cell of width a and length b. However, in the limit in which $a \ll b$ one finds theoretically [531] that

$$q = \text{constant}\ b \log a \ , \tag{17.2}$$

in agreement with Fig. 17.16.

Although this result can be obtained analytically either by contour transformations or by simpler approximations, theoretical physicists may recognize it more directly as the result of a dimensionality argument. As one dimension goes to zero, it can enter the expressions depending upon it at most logarithmically (zero-dimensionality). Therefore the capacitance of a nearly one-dimensional rectangular parallel-plate condenser must go as $b \log a$.

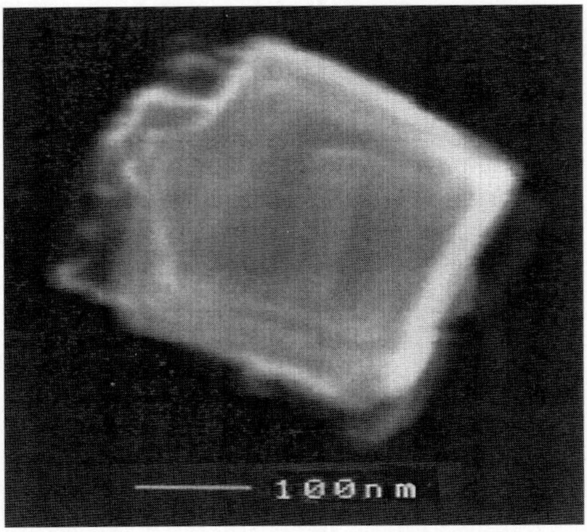

Fig. 17.17. Bi_2O_3 nanoelectrode crystallizing on the surface of bismuth titanate PLD film

17.5 Bismuth Titanate

$Bi_4Ti_3O_{12}$ (BiT) is of somewhat less interest than SBT for memory devices, but its properties are similar, and more detailed physical studies have been made on its nanoscale properties. Recently, BiT with La has been shown to be a good RAM material [532]. Figure 17.17 shows a single crystal of 150 nm grown on the surface of BiT. Careful inspection reveals the spiral growth rings. Figure 17.18 shows that these nanoscale electrodes are self-patterning on the BiT surface, with cross sections revealing an average height of 45 nm.

In Fig. 17.19 we see the radial correlation function for neighboring nanoelectrodes; and the inset in this figure shows that the angular correlation is strongly peaked every 90°, which Fig. 17.3 reveals to be along the underlying Si [100] axes (probably due to strain orientation).

Of great interest is the leakage current of these δ-Bi_2O_3 nanoelectrodes on BiT. Figure 17.5 shows $J(V)$ from +4 to −4 V. The inset is a plot of $J^{1/2}(V)$ versus V, illustrating perfect quadratic (space charge limited) currents.

Cross-sectional TEM measurements on this system (Fig. 17.20) reveal the (111) orientation of the mesa-like δ-Bi_2O_3 nanoelectrodes, rather more clearly in this case than in the analogous studies of Bi_2O_3/SBT.

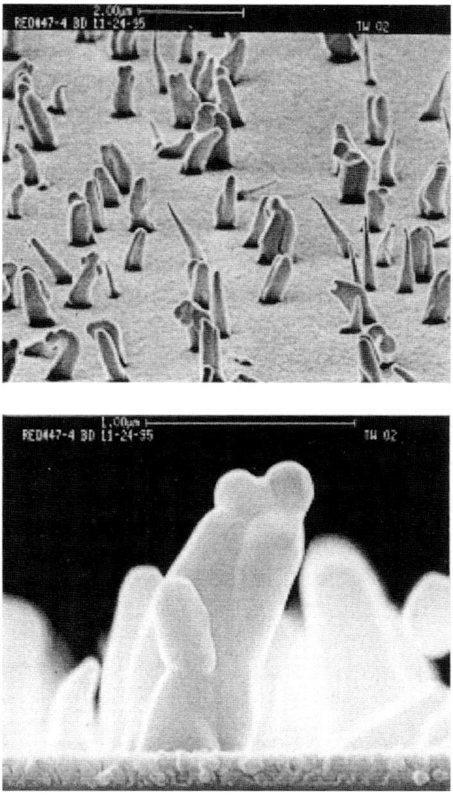

Fig. 17.18. Self-patterning of bismuth oxide nanoelectrodes on bismuth titanate films: (*above*) plan view (SEM), (*below*) cross section (AFM)

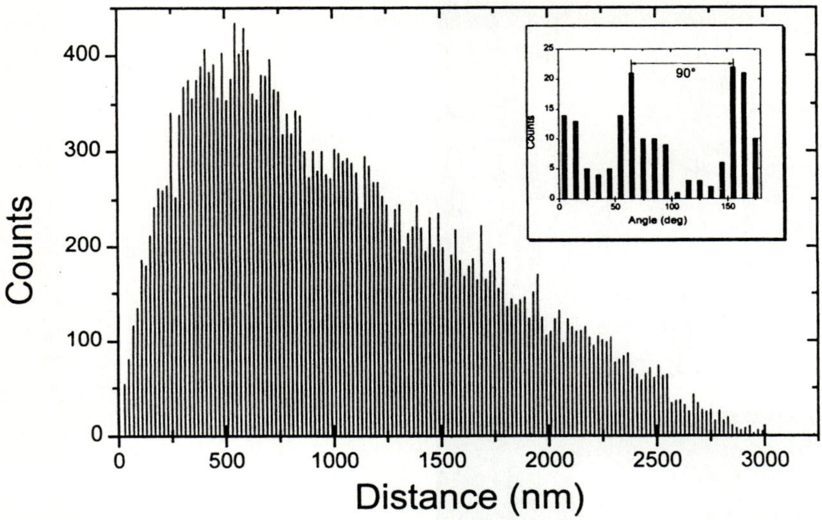

Fig. 17.19. Radial and angular (*inset*) correlation functions for neighboring nanoelectrodes of bismuth oxide on bismuth titanate

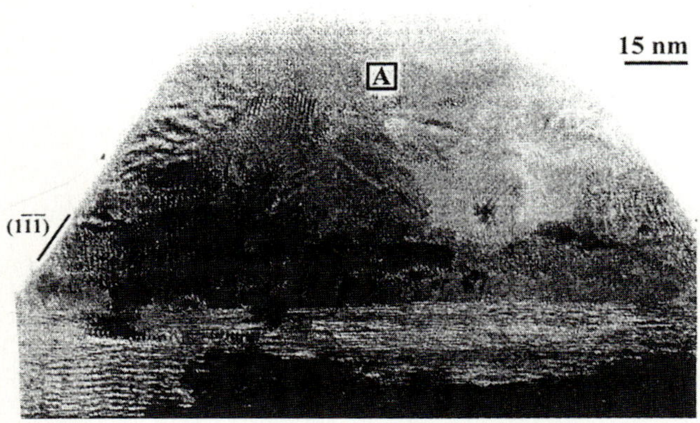

Fig. 17.20. TEM cross section showing (111) lattice orientation of δ-Bi_2O_3 nanoelectrodes

18. Drawbacks and Disadvantages

The main difficulties associated with nonvolatile ferroelectric memories are listed below. Until these are overcome, it will be difficult for NV-RAMs using ferroelectric thin films to displace flash EEPROMs for most applications.

18.1 High Processing Temperatures

Ferroelectric thin films of the PZT family require a minimum of 723 K to convert fully into the perovskite phase from pyrochlore. This is inconvenient but quite feasible. More difficult are the temperatures > 873 K required to convert SBT films from fluorite to Aurivillius phase. Such a high temperature eliminates the use of aluminium metallization and makes very difficult double-metal devices. (Most prototype NV-RAMs reported in the literature employed only single-level metal, which is not suitable for sophisticated devices.) In addition, high processing temperatures for SBT generate several other problems: a bismuth diffusion barrier must be designed to prevent bismuth from getting into the bottom electrode and underlying silicon IC; a top electrode material must be developed that is compatible with forming gas anneal at high temperatures; and the reactivity of bismuth with platinum at high temperatures must be kept in mind. Rhodium–palladium alloys seem useful in these applications.

18.2 Toxicity

It may seem frivolous to worry about the toxicity of lead in a semiconductor fabrication facility (FAB) that uses deadly arsine or phosphine, but in fact the Environmental Protection Agency (EPA) in the USA has increasing tight restrictions on lead in the workplace – so much so that IBM has been considering replacement of all its lead-based solders by bismuth compounds. Both worker safety and hazardous waste disposal issues therefore favour SBT-family ferroelectrics over PZT. Bismuth, in the form of bismuth oxide, is edible and in fact is the major ingredient in non-perscription drugs such as Pepto-Bismal, used for diarrhoea and stomach upset.

18.3 Surface and Interface Phenomena

The defect concentrations in ferroelectric oxides are typically a few times 10^{20} cm^{-3} near the surfaces. This is more than a thousand times that of a good silicon chip. Many of these defects are 'native defects', such as oxygen vacancies; therefore, they cannot be greatly lowered by making the materials purer and relatively free of elemental contaminants. Such defect densities result in very large charge injection J' from surface traps when electric fields are applied to switch the devices. These injected charges usually mask the displacement vector D that the sense amplifiers in the memories need to measure. This problem arises whenever the ferroelectric is in direct contact with the silicon (or GaAs) surface, as in a ferroelectric-gated FET. It is not important in passgate NV-RAM architecture (or in a DRAM) where the ferroelectric contacts only a metal electrode.

18.4 Half-Select Disturb Pulses

The fact that ferroelectric materials exhibit activation field dependent switching rather than true voltage thresholds means that it is still not practical to try to eliminate the passgate transistors and design raw-array memories.

18.5 Fatigue

The use of oxide electrodes has significantly lessened the problem of fatigue in PZT NV-FeRAMs. The reasons for this are still under discussion, but the primary cause is probably a reduction of oxygen diffusion from the ferroelectric into the electrode. Sanchez et al. showed as early as 1991 [534] that diffusion of oxygen into Pt bottom electrodes was much less than diffusion of oxygen into Au top electrodes, resulting in an electrical asymmetry that worsened with time and repetitive bipolar cycles. Similarly, diffusion of material from electrodes into the dielectric is also a problem. Fatigue was never a problem for SBT devices. Lower operating voltages (3.3 V or 1.1 V compared with 5.0 V) in the near future will also reduce fatigue substantially. (However, operating a PZT memory at 1.1 V will be very difficult for reasons of coercive voltage.)

18.6 Imprint

Imprint problems exist with all ferroelectric films at present. Imprint can be minimized in two ways: from a materials point of view, charged impurities must be reduced in concentration to avoid electret effects. From an electrical engineering point of view, it may be useful to program in restore (or 'rejuvenation') pulses periodically (e.g., every million cycles).

18.7 Scaling

There is no rigorous, detailed theory of finite size effects either for film thicknesses or edge effects (capacitor area effects). As the nanoscale of 0.1 microns is reached in the next year or so, this needs to be done.

18.8 Short-Term Current Transients

At times in the μs regime, the switched charge sensed in ferroelectric switching films is more relaxational P/τ than true displacement current D from switched polarization $\mathrm{d}P/\mathrm{d}t$. This will complicate the sensing of very fast memory devices.

The short time current transients measured by Dietz et al. [535] generally satisfy a power law with a time exponent of about -0.8. This is too large to be due to Cottrell desorption, which gives -0.5 exactly [535], but it is very close to the value for ionic conduction found by Scott et al. [536]. This correlates very well with the recent study of space charge limited ionic transients by Zafar et al. [537] and suggests that the currents in the 10 μs to 1.0 ms regime with such exponents are ionic in ferroelectric films; this in turn is related to electret effects.

18.9 Breakdown

The mechanisms for breakdown are reviewed at some length in this book. However, there is still no comprehensive understanding of this process in most ferroelectric films. The initiation stage seems to be avalanche, with the intermediate time scale characterized by impulse thermal breakdown. Avalanche initiated by impact ionization has been a contentious hypothesis in II–VIs for forty years [538], with the RCA group seeing breakdown due to field emission but not impact ionization across the bandgap Therefore the exact physical mechanism of breakdown in ferroelectric oxides will require much more careful study. Longer time processes may involve dendrite-like growth of microscopic conducting pathways, particularly on oxide electrodes. These form a potential 'umbrella' that prevents large areas of the capacitors from switching; and they eventially cause electrical shorts. These processes will be most important in DRAM films, where maximum capacitance and chip bit-density will require ferroelectric films of order $d = 20$ nm thick. (Thinner films will probably not be used even if they can be made pinhole-free, because Fowler–Nordheim tunneling will produce unacceptably large leakage currents.)

18.10 Leakage Currents

Leakage current is not a problem at present for SBT or PZT films 200 nm thick. The use of compound electrodes (e.g., Pt plus Ir/IrO_2) seems a satisfactory solution for PZT. Present NV-FeRAMs achieve a leakage of 100 nA/cm^2 at 3 V. However, leakage currents are still problems for BST DRAMs, simply because the desired thickness is about an order of magnitude less (20 nm). A combination of the best deposition (MOCVD) for BST with good, well-annealed Pt electrodes is the best solution at present for BST DRAMs. Modification of BST/Pt depletion widths via acceptor or donor doping, as shown by Lee and Park in Fig. 18.1, is the most promising scheme for reducing BST leakage currents.

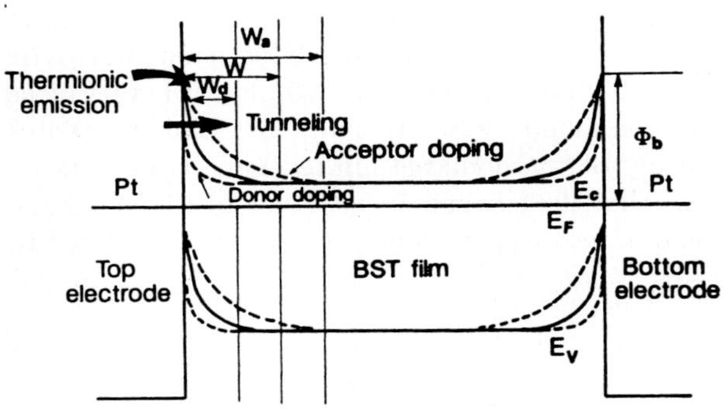

Fig. 18.1. Modification of BST/Pt depletion widths via acceptor and donor doping. After Lee and Park [533]

18.11 Radiation Hardness

Radiation effects in bulk ferroelectrics have been studied for more than forty years; in ferroelectric thin films, only recently.

In 1957 Wittels and Sherill showed [539] that large total doses of X-rays converted tetragonal, ferroelectric barium titanate into the cubic paraelectric phase. Lefkowitz and Mitsui later found [539], surprisingly, that irradiation actually lowered the coercive field of this material, whereas in other ferroelectrics, such as Tri-Glycine Sulfate (TGS) it increases it [540], and a large internal bias field develops. For prolonged exposure in TGS the bias field exceeds the coercive field, so that switching occurs for only one polarity of applied voltage. The remanent polarization P_r generally decreases with irradiation, often to zero [541], so that no hysteresis loops can be observed. This is usually due to pinning of domain walls by defects created by the irradiation [542].

Our tests on PZT used 5 Mrad total dose [543] and 10^{11} rad/s dose rate. We found that under these conditions E_c increased only about 2.5% and the bias voltage actually improved (decreased) after irradiation (probably due to induced displacements of charged defects). Spontaneous polarization decreased an acceptable 6%. However, radiation damage was more severe in cells stored in shorted conditions (as opposed to open-circuit conditions).

Ferroelectric RAMs have excellent radiation hardness compared with DRAMs or SRAMs. They are generally immune to Single-Event Upset (SEU) and withstand very high dose rates and total dosages of both X-rays and neutron flux. The radiation hardness testing of ferroelectric thin-film random access memories was probably first carried out by Rockwell on some KNO_3 memories made by George Rohrer. The first studies on PZT were by the present author [543]. A 2 MeV Febetron 705 X-ray unit was employed, utilizing a Field Emission Corp. 545 A e-beam tube. Total dose analysis was performed with Cs^{137} calibrated LiF dosimeters. It was found that PZT memories would withstand 5 Mrad of gamma rays with full functionality and small decrease in switched charge, but that the information retention was best under exposure if the cells were stored in an open-circuit configuration. More detailed studies were done subsequently on PZT at Sandia [544] using samples from National Semiconductor Corp. The results were similar except that one batch of PZT from National Semiconductor failed the tests, showing extreme degradation at 1 Mrad. The reasons for this inferior PZT batch were never clarified. Very recently the radiation hardness of PZT irradiated with oxygen ions (around 10^{13} cm^{-2} total dose) was also studied [544], with good results. The radiation hardness of FeRAMs is a strong argument for their use in NASA and commercial satellite computer memories, which must pass continuously through the Van Allen belt. For these purposes, radiation-hard FeRAMs can replace miniaturized 'core' memory. Although both are nonvolatile and radiation-hard, the ferroelectric devices are lighter in weight and use less power than does core memory.

18.12 Future

In the immediate future (next three years) the commercial product line will include low density nonvolatile FeRAMs, mostly with SBT and possibly with PZT, for smart cards and embedded memories. But the greater impact will be the use of ferroelectric capacitors in DRAMs. Figures 18.2–18.4 present the state-of-the-art ferroelectric DRAM technology (NEC). Figure 18.2 shows the layout of word-line, bit-line, and storage nodes for a 0.14 μm gate width RAM, for which the cross section is given in Figs. 18.3 and 18.4. This is a tungsten silicide metallized BST DRAM with a very thin 26 nm BST film. The top and bottom electrodes are both elemental Ru, but the bottom has a thick ruthenium oxide–ruthenium sandwich structure cemented with TiN onto a polysilicon plug. Similar structures for SBT are also already developed

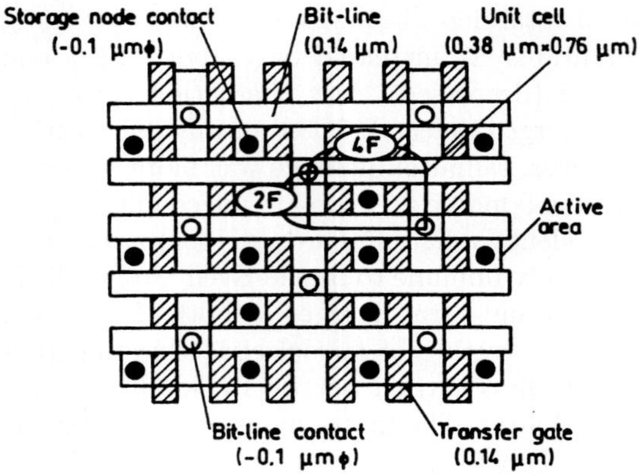

Fig. 18.2. 0.14 µm gate width DRAM schematic from NEC, showing 0.1 µm diameter storage node and bit-line contact, and $0.38 \times 0.76\ \mu m^2$ cell area

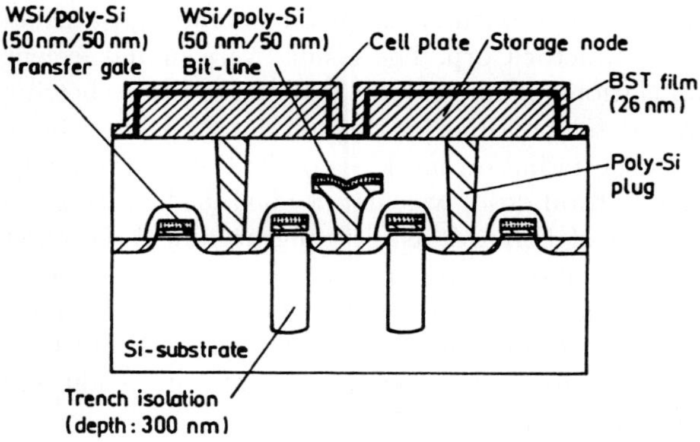

Fig. 18.3. BST DRAM design with tungsten silicide transfer gate and bit-line, polysilicon plug, and 300 nm trench (NEC)

and emphasize the bottom electrode diffusion barrier to prevent Bi diffusion onto the Si substrate. This diffusion barrier is generally proprietary yet, but platinum–rhodium alloys have been discussed in this context.

It is well to reflect upon the past before predicting the future. In 1978 Taylor reviewed [545] very well the original work on semiconductor devices [546, 547, 548, 549, 550]. These had thick ferroelectrics and thin semiconductor films. It has required twenty years of improved deposition techniques for oxides, some of which were acquired through the international popularity of high-temperature superconducting oxides, to reverse this geometry and use

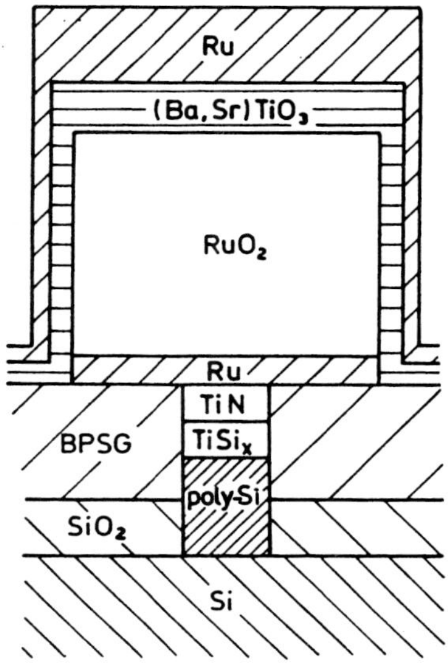

Fig. 18.4. NEC ruthenium/ruthenium oxide metallization for BST DRAM with titanium nitride/titanium silicide/polysilicon plug for high density vertical integration

thin ferroelectric films on thick semiconductors (Si or GaAs). The future will also benefit from breakthroughs in deposition. Although some of these are unpredictable, it seems likely that nanophase techniques will be involved and that self-assembly may be, too.

A. Exercises

1. Estimate the change in coercive field in an SBT nonvolatile memory between the 60 Hz Sawyer–Tower measured value and actual operation at (a) 100 MHz, (b) at 300 Mhz.

2. Assuming reasonable $C(V)$ values for BST calculate the dielectric loss in BST capacitors at 3.0 V bias at 15 GHz from Tagantsev's intrinsic loss model, assuming a zero-bias value of $Q = 50$ and $Q = 240$ at 1.0 V bias.

3. Summarize the evidence supporting: (a) avalanche breakdown, (b) impulse thermal breakdown, (c) dc thermal breakdown, and (d) Gerson–Marshall connected-defect breakdown in PZT and in BST.

4. Below 400 kV/cm what leakage mechanism is dominant in BST? How can you distinguish between bulk Poole–Frenkel and interface-limited Schottky currents in BST?

5. From 400 kV/cm to 2 MV/cm, what leakage current mechanism is dominant in BST?

6. Summarize the evidence for and against space charge limited currents in ferroelectric thin films.

7. Estimate the ambipolar diffusion width in PZT/Pt.

8. Is negative resistivity in ferroelectric thin films (PZT or SBTN) due to double injection or trap filling/emptying? What evidence supports your view?

9. What evidence exists for double injection (holes and electrons) in ferroelectric thin films in general?

10. What is a 'shadow RAM'? Why is it a conservative device? What are its disadvantages?

11. What is the advantage of a 2T–2C ferroelectric RAM cell? What are the disadvantages?

12. What non-ferroelectric devices or materials exhibit electrical hysteresis? Why?

13. Summarize any evidence from the literature for supersonic domain wall velocities in ferroelectrics. We know that magnetic domain walls can be supersonic. Why are magnetic and ferroelectric domain walls qualitatively different? (A term paper topic?)

14. What is the primary problem in producing a ferroelectric FET, e.g., with a PZT gate?

15. What is a typical coercive field in a ferroelectric thin film? Why is 10 kV/cm a plausible lower limit that might be achieved in PZT?

16. If I increase the voltage from 2 V to 10 V in a ferroelectric thin film, by what percentage does the switching time decrease?

17. If I used Cu electrodes on PZT, what breakdown field would be expected?

18. If I decrease the capacitor area to 100 square microns from 10 000, how much faster does it switch? By how much does the breakdown field increase? Why?

19. If I increase the applied field by 10 times in a BST film, how much faster does breakdown occur?

20. How do domain wall velocities vary with voltage?

21. The load resistors in early potassium nitrate thin film memories at Ramtron were 1 kΩ. Estimate the minimum RC-limited switching time for a 100 × 100 micron capacitor 300 nm thick with a dielectric constant of 17. Make the same estimates for an identical PZT film with $\varepsilon = 1300$.

22. Assuming that PZT with grain size averaging 1 micron has domains 0.05 microns in length, estimate the domain size of PZT grains 0.4 microns in diameter.

23. Estimate the breakdown field in PZT, using impulse thermal breakdown. How is this influenced by depletion near grain boundaries? Would a single-crystal PZT film have a higher or lower breakdown field? Why?

24. Summarize the evidence for n-type inversion layers in nominally p-type PZT or BST.

25. Estimate the change in work function for a 10% sub-monolayer coverage of Pt with oxygen. Using the Von Hippel relationship, estimate the change in breakdown field this would cause for PZT/Pt. Estimate the number of fatigue cycles required to liberate this much oxygen in a 300 nm thick PZT film. What did you assume about charge injection rates? Can this explain the increase in leakage current with fatigue observed in PZT/Pt?

26. Using the nucleation rate of $1.5 \times 10^{22}\ \mathrm{s^{-1}m^{-2}}$ on PZT surfaces, estimate the number of domains that can nucleate in a single 40 nm grain of PZT during a 100 ns voltage pulse.

27. What are the three different regimes exhibited in switched charge versus switching cycle fatigue data? Within the theory of Duiker, why do they occur?

28. By how much must I vary the voltage applied to a 250 nm thick PZT film to compensate for an increase in temperature from ambient to 200°C and still maintain exactly the same switching speed?

29. Why is the Cottrell desorption of oxygen from the Pt electrodes unlikely to account for the $t^{-0.75}$ current transients observed in PZT/Pt devices?

30. Given an activation field dependence with activation field 100 kV/cm, how much faster is a PZT memory at 1 MV/cm than at 300 kV/cm?

31. If I make a new ferroelectric by adding an ingredient to PZT, such that the spontaneous polarization is halved, what is the likely activation field for switching in the new material? (Hint: empirical theory of Glass.)

32. $C(V)$ often varies in ferroelectric films as $V^{-1/n}$. Discuss values of n from different models. What might you infer from experimental values of $0.3 < n < 0.4$?

33. What is the space charge dependence upon depth z as one moves a distance into a ferroelectric film from a metal electrode? Is it different for ohmic contacts compared with Schottky blocking contacts?

34. Calculate the effective exponent in the thickness dependence of a space charge limited current device for an ambipolar diffusion width of 20% of the film thickness. Does this diffusion also introduce a change in the voltage exponent? Why?

35. Use the Rose relationship to estimate the density of traps within $k_B T$ of the Fermi energy in PZT if the dielectric constant is 1300 and the leakage current is 3 $\mu A/cm^2$ at 4 V and 6 $\mu A/cm^2$ at 12 V.

36. Using Okazaki's relationship, estimate the difference in switching speeds for + to − and for − to + switching in PZT having a space charge field of +5.0 kV/cm.

37. Using Arlt's relationship, $E_{bias} = \Delta n \Delta W / P_r$, estimate the number density of charged defects Δn that would be required to produce the field in the preceding problem, if the energy difference ΔW for such dipolar defects in fields $E = +400$ kV/cm and -400 kV/cm is pE, with $p = 4e \times 0.01$ nm ($e =$ electron charge). (Don't forget the factor of ε_0 in SI units.)

38. The best lithium-niobate gated FET has a gain of about 55? How is this estimated? What do you think limits the gain?

39. Use the Rosenman relationship, $E = p' \Delta T [(\varepsilon g/d) + 1]^{-1}$, to estimate the field E generated in a PZT pyroelectric detector if the film thickness d is 500 nm and the gap separation g between target and microchannel plate is 10 microns. Here p' is the pryoelectric coefficient.

40. Using data in the text figures, estimate the minimum access time in a PZT memory if the sense amplifiers require a 2:1 signal ratio to discriminate unambiguously between switching and non-switching current pulses. How does your numerical access time estimate change if the sense amplifiers actually require not a signal ratio but an absolute difference of 100 mV?

41. What is the appropriate fringing field correction of the capacitance of a ferroelectric capacitor 0.7 microns on a side and 300 nm thick? (Hint: text by Smyth.) Where do the lines of force go in such a device?

42. What is the maximum phase shift one can obtain at 50 V across a 5 micron wide capacitor (interdigital surface electrodes, transverse field) in an RC-circuit with a 52 ohm coaxial line and a strontium titanate capacitor of 500 pF at $V = 0$ (dielectric constant of 310)?

43. The figure of merit for lead scandium tantalate (PST) pyroelectric detectors is 5 times that of PZT, yet the performance of PZT in integrated Si pyroelectric detector devices is the same as for PST. Why?

44. Some prototype ferroelectric FETs exhibit clockwise electric hysteresis loops. Others exhibit counter-clockwise loops. Why? Which ones are actually functioning as ferroelectric FETs?

45. The minimum processing temperature of PZT memory films is 450°C. Why? What choice of metallization does this permit?

46. The minimum processing temperature for SBT memory films is about 550°C. Why? What choice of metallization does this permit?

47. Why should SBT films be prevented from contact to Pt electrodes at high temperatures?

48. What are the advantages of CVD over sputtering for ferroelectric films? What are the main disadvantages of laser ablation? How can you carry out CVD with viscous precursors having low volatility and low vapour pressures (as in SBT)?

49. How can you most easily measure the droplet size for 'mist' deposition?

50. Silver paint electrodes usually produce breakdown at lower fields than does sputtered silver. Why?

51. Sputtered silver electrodes produce breakdown at lower fields than does sputtered platinum. Why?

52. Oxide electrodes often produce dendritic shorts. Why?

53. DRAM designers require 30 fF of capacitance per cell. If I can make an SBTN cell 45 nm thick with a dielectric constant of 400, can I achieve this without trenching in a 1 Gbit DRAM with 0.5×0.5 micron capacitors?

54. The capacitance of a ferroelectric drops rapidly as field is increased. If I design a DRAM assuming that the zero-voltage capacitance ($\varepsilon = 300$) is maintained throughout the applied voltage pulse duration, will I have a serious problem?

55. Maximum DRAM leakage current toleration is between 1 and 10 $\mu A/cm^2$. Can I use 25 nm thick PZT? Can I use 45 nm thick BST? Can I use 45 nm thick SBT? (Assume a 2.5–3.3 V design.)

56. The NEC 1 Mbit ferroelectric RAM has an access time of 60 ns. The Matsushita 256 kbit ferroelectric RAM has an access time of 100 ns. How much of this is intrinsic to the ferroelectric (domain speed limited)? How much of it arises from the load resistance? Could you make a 900 ps ferroelectric memory? How? Or why not?

57. What is the figure of merit for a ferroelectric nonvolatile RAM? If the dielectric constant should be low, why use PZT with a value of 1300 instead of potassium nitrate with a value of 17? If a high spontaneous polarization and low dielectric constant are optimal, which is best: barium magnesium fluoride, potassium nitrate, or PZT?

58. From a practical point of view, the coercive field for a ferroelectric RAM should be < 100 kV/cm. Why? What are the device design problems if it is 200 kV/cm? What are the processing problems if it is 200 kV/cm?

59. Langmuir–Blodgett films of TFE-PVF2 1.0 nm thick have demonstrated ferroelectric switching. What are the practical problems in making a nonvolatile RAM from them?

60. When I shine light on PZT its coercive field changes significantly due to photovoltaic effects. Can I use this to record a photograph on a megapixel PZT array without using a pyroelectric effect? How could I make grey scales? Can I use this to store X-ray images digitally without film?

61. The torque generated in an electrostatic micromachine is proportional to the dielectric constant of the thin-film cog. What mechanical advantage is obtained via fabrication with $SrTiO_3$ instead of SiO_2?

62. The emission of electrons from the surface of PZT can generate 1 kA/cm^2 in synchronized bursts with 1 ns rise times. Can you use this to make electron sources for accelerators? For flat-panel displays? What limits the current density of electrons emitted from PZT surfaces under applied voltages? What limits the rise time?

63. A theoretical publication in 1973 claimed that ferroelectricity in films thinner than 0.4 microns would be destroyed by depolarization fields. Yet ferroelectric films as thin as 0.0009 microns have been switched. How thin can one make a ferroelectric film without destroying or destabilizing the ferroelectricity?

64. What is the minimum domain size in a ferroelectric film? What is the minimum domain wall width in a ferroelectric film? Are the two numbers related?

65. If I increase the conductivity in a ferroelectric film, what happens to the asymmetry in the switching time? (Takahashi model.)

66. RAM designers think that 1 or 2 μC/cm^2 is an absolute minimum for remanent polarization in a ferroelectric RAM. Why?

67. In GASH (Guanidinium Aluminum Sulfate Hexahydrate) the nucleation time is longer than the domain growth time. TGS is similar. Would that make them good or bad ferroelectric memories?

68. Some of my friends, experts on silicon microelectronic devices, would like to do 'accelerated stress tests' on ferroelectric memories by subjecting them to voltage pulses at elevated temperatures. Why is this not a valid 'accelerated' test procedure?

69. What is the difference between true epitaxial films and grain oriented films? Does it affect their switching properties?

70. When I do Auger spectroscopy of BST, I find that much of the Ti is 3+ rather than 4+. Is this an artifact due to Auger bombardment, or does it reflect the true state of Ti in the ceramic?

71. How hot does the ferroelectric–electrode interface get in PZT/Pt when 1 $\mu A/cm^2$ passes through it? What is the peak temperature? How does temperature T fall off with distance as one moves into the ferroelectric film? That is, how large is the temperature gradient? How would you measure this temperature profile? Does this not make the use of Schottky models with constant temperatures a worry?

72. Are the Peltier, Seebeck, or Thomson effects measurable in these PZT/Pt devices? If so, why are they ignored in models for $J(V)$?

73. If the ferroelectric–electrode interface gets up to 500 K, what is the pyroelectric voltage generated in PZT/Pt? Should that voltage be inserted as a correction to the Schottky equation used to model $J(V)$ data?

74. What is the electron mobility in PZT? In BST? What is the hole mobility? If these are p-type wide bandgap semiconductors in bulk, why are the only carriers in thin-film ferroelectrics electrons and not holes? Why do we not see visually light emitted from recombination processes in PZT with injected currents of electrons?

75. Li and Lu claim that about half of their barium titanate films have Poole–Frenkel conduction and half have Schottky conduction. How can you distinguish the two experimentally? Which will be dominant in films with large surface trap densities?

76. What are the principal surface donor states in BST films? Do they pin the metal electrode Fermi energies at the electrode interface? What are the trap depths (pinning energies)?

Bibliography

Chapter 1

1. Jona F. and Shirane G., *Ferroelectric Crystals* (Pergamon, Oxford, 1962)
2. Fatuzzo E. and Merz W. J. (North-Holland, Amsterdam, 1967)
3. Mitsui T., Tatsuzaki I., and Nakamura E., *An Introduction to the Physics of Ferroelectrics* (Maki-Shoten, Tokyo, 1969; English translation: Gordon & Breach, London, 1976)
4. Megaw H. D., Acta Cryst. **7**, 187 (1954)
5. Lines M. and Glass A. M., *Principles and Applications of Ferroelectrics and Related Materials* (Clarendon, Oxford, 1977)
6. Xu Y., Chen C. J., Xu R., and MacKenzie J. D., Ferroelec. **108**, 1653 (1990)
7. Sessler G. M., *Electrets* (Springer-Verlag, New York, 1987)
8. Su Q., Rabson T. A., and Robert M., Integ. Ferroelec. **18**, 415 (1997)
9. A recent review of imprint is Benedetto J. M., Integ. Ferroelec. **15**, 29 (1997)
10. Landau L. D., Phys. Z. Sowjetunion **11**, 26 (1937)
11. Devonshire A. F., Phil. Mag. **740**, 1040 (1949); **742**, 1065 (1951)
12. de Gennes P. G., Sol. St. Commun. **1**, 132 (1963)
13. Blinc R., Proc. Int. Course on the Theory of Condensed Matter (IAEA, Vienna, 1968), p.395
14. Nelmes R., *Ferroelectrics* (in press)
15. Peercy P. S., Light Scattering in Solids, eds. M. Balkanski, R. C. C. Leite, and S. P. S. Porto (Flammarion, Paris, 1976), p.782
16. Kozlov G. V., Volkov A. A., Scott J. F., Feldkamp G. E., and Petzelt J., Phys. Rev. **B28**, 255 (1983)
17. Windsch W. and Volkel G., Ferroelec. **24**, 195 (1980)
18. Chen T. and Scott J. F., Phys. Rev. **40**, 8978 (1989)
19. Scott J. F. and Chen T., Phase Transitions **32**, 235 (1991)
20. Scott J. F. and Chen T., J. Raman Spec. **21**, 761 (1990)
21. Errandonea G., Phys. Rev. **B21**, 5221 (1980); Scott J. F., Ferroelec. **166**, 95 (1995)
22. Greer A. L., Habbal F., Scott J. F., and Takahashi T., J. Chem Phys.**73**, 5833 (1980)
23. Scott J. F., Habbal F., and Zvirgzds J. A., J. Chem. Phys.**72**, 2760 (1980);**69**, 4984 (1978)
24. Cochran W., Adv. Phys. **9**, 387 (1960); **10**, 401 (1961); Scott J. F., Rev. Mod. Phys. **48**, 83 (1975); Jaynes E. T., PhD thesis, Princeton (1950); Jaynes E. T., *Ferroelectricity* (Princeton University Press, 1953); Wigner E. P., private communication cited by Jaynes; Kristoffel N. and Konsin P., Phys. Stat. Sol. **21**, K39 (1967); **28**, 731 (1968); Sov. Phys. Sol. St. **13**, 2113 (1972); Ferroelectrics **6**, 3 (1973)

25. Griffiths R. B., J. Chem. Phys. **43**, 1958 (1965); Scott J. F., J. Phys. Soc. Jpn. **58**, 4487 (1989)
26. Rushbrooke G. S., J. Chem. Phys. **39**, 842 (1963); **43**, 3439 (1965)
27. Scott J. F., J. Phys. Chem. Sol. **57**, 1439 (1996)
28. Lopez-Echarri A., Tello M. J., and Gili P., Sol. St. Commun. **36**, 1021 (1980)
29. Reese W., Sol. St. Commun. **7**, 969 (1969)
30. Scott J. F., Habbal F., and Hidaka M., Phys. Rev. **B25**, 1805 (1982)
31. Toledano P. and Toledano J.-C., *The Landau Theory of Phase Transitions* (World Scientific, Singapore, 1987)
32. Pompe W., Speck J. S., et al., J. Appl. Phys. **83**, 2742 (1998); **83**, 2754 (1998); **78**, 1696 (1995); **79**, 4037 (1996); Takashige M., Hamazaki S.-I., Fukurai N., and Shimizu F., J. Phys. Soc. Jpn. **6**, 1848 (1997); Zybill C. E., Koch F., Wersing W., et al. (in press 1999). For earlier AFM studies of surface domains in ferroelectric films, see Gruverman A. et al., Nanotechnology **8**, A38 (1997); Appl. Phys. Lett. **69**, 3191 (1996); and Hidaka T. et al., Appl. Phys. Lett. **68**, 2358 (1996). Pertsev N. A., Zembiligotov A., and Tagantsev A. K., Ferroelectrics **223**, 79 (1999)
33. Yamada Y. and Hamaya N., J. Phys. Soc. Jpn. **52**, 3466 (1983)
34. Viehland D., private communication (1995); Viehland D., Forst D., and Li J.-F. (in press)
35. Sinharoy S., et al., J. Vac. Sci. Tech. **A9**, 409 (1991)
36. Scott J. F., Rep. Prog. Phys. **42**, 1055 (1979)
37. Batra I. P. and Silverman B. D., Sol. St. Commun. **11**, 291 (1972)
38. Landauer R., private communication (1987)
39. Mills D. L., Phys. Rev. **B3**, 3887 (1971)
40. Lubensky T. C. and Rubin M. H., Phys. Rev. **12**, 3885 (1975)
41. Tilley D. R. and Zeks B., Sol. St. Commun. **49**, 823 (1984)
42. Zhong W. L., Wang Y. G., and Zhang P. L., Ferroelec. Rev. **1**, 131 (1998) give an up-dated comprehensive review; see also Qu B., Zhong W., Wang K., and Wang Z., Integ. Ferroelec. **3**, 7 (1993)
43. Scott J. F. et al., Physica **B150**, 160 (1988)
44. Bune A. V., Fridkin V. M., Ducharme S., et al., Nature **391**, 874 (1998). There may be a relationship between the domain-free switching in ferroelectric films reported by Bune et al. and the domain-free magnets reported by Stamm et al., Science **282**, 449 (1998). Stamm et al. find that the demagnetization field dominates above a critical thickness d^* and exchange energy dominates below this size. d^* is about 10 monolayers
45. Chew K.-H., Osman J., Stamps R. L., Tilley D. R., and Webb J. F., Integ. Ferroelec. **23** (in press 1999)
46. Katiyar R. S. (private communication)
47. Watanabe Y. and Sawamura D., Jpn. J. Appl. Phys.**36**, 153 (1997)
48. Ishikawa K. et al., Phys. Rev. **151**, 378 (1966)
49. Uchino K., Sadanaga E., and Hirose T., J. Am. Ceram. Soc. **72**, 1555 (1989); Ferroelec. Lett. **44**, 55 (1982)
50. Gachigi K., Kumar U., and Dougherty J. P., Ferroelec. **143**, 229 (1993)
51. Payne D. A., Phys. Rev. **B54**, 3158 (1996); Frey M. H. and Payne D. A., Ferroelec. (1998 in press); Randall C. A., McCauley D. E., and Cann D. P., Ferroelec. (1998); Cao W. and Randall C. A., J. Chem. Phys. Sol. **57**, 1499 (1996)
52. Tanaka M. and Makino Y., Ferroelec. Lett. **24**, 13 (1998); Tang X. G., Guo H. K., Zhou Q. F., and Zhang J. X., NanoStruc. Mater. **10**, 161 (1998)

Chapter 2

53. Scott J. F. and Araujo C. A., Science **246**, 1400 (1989)
54. Schubring N. W., Dork R. A., and Nolta J. P., J. Appl. Phys. **38**, 1671 (1967)
55. Schubring N. W., Dork R. A., and Nolta J. P., *Ferroelectricity*, ed. E. F. Weller (Elsevier, Amsterdam, 1967) p.269
56. Scott J. F. et al., J. Appl. Phys. **62**, 4510 (1987)
57. Dimmler K., et al., J. Appl. Phys. **61**, 5467 (1987)
58. Ramesh R. et al., Integ. Ferroelec.**1**, 1 (1992)
59. Sumi T. et al. (Matsushita–Panasonic), ISSCC Digest, paper FA16.2, p.68 (1994)
60. Sumi T. et al., Integ. Ferroelec. **6**, 1 (1995)
61. Sumi T. et al., Jpn. J. Appl. Phys. **35**, 1516 (1996)
62. Koike H. et al. (NEC), Proc. ISSCC, 16 Feb 1995; reproduced by Paz de Araujo C. A., Scott J. F. and Taylor G. W., *Ferroelectric Thin Films* (Gordon & Breach, New York, 1996) p.4
63. Fazan P. C. (Micron), Integ. Ferroelec. **4**, 247 (1994)
64. Gnade B., Summerfelt S. R., and Crenshaw D. (Texas Instruments), *Science and Technology of Electroceramic Thin Films*: NATO ASI Series, eds. Auciello O. and Waser R. (Kluwer, Dordrecht, 1995) p.284
65. Kotecki D. E. (IBM), Integ. Ferroelec. **16**, 1 (1997); Bernacki S. et al., Integ. Ferroelec. **3**, 1 (1993)
66. Kinney W., Integ. Ferroelec. **4**, 131 (1994)
67. Fukushima T. et al., *VLSI Circuit Digest of Technical Papers* (1996) p.46
68. Duiker H. M. et al., J. Appl. Phys. **68** 5783 (1990)
69. Song T. K., Ahn J., Yang B., Aggarwal S., and Ramesh R., J. Korean Phys. Soc. **32**, S1721 (1998); Wouters D. J. et al., J. Phys. IV France **8**, 205 (1998)
70. Okazaki K. and Nagata K., J. Am. Ceram. Soc. **56**, 82 (1973)
71. Okazaki K. et al., Ferroelec. 7, 153 (1974)
72. Lang S. B. and Das-Gupta D. K., J. Appl. Phys. **59**, 2151 (1986)
73. Cheng Z. and Yao X., Ferroelec. **109**, 155 (1990)
74. Brennan C. Integ. Ferroelec. **8**, 335 (1995)
75. Brennan C., Integ. Ferroelec. **8**, 93 (1995)
76. Melnick B. M., Cuchiaro J. D., McMillan L. D., Paz de Araujo C. A., and Scott J. F., Ferroelec. **112**, 329 (1990)
77. Maderic B. P., Sanchez L. E., and Wu S.-Y., Ferroelec. **116**, 65 (1991)
78. Raleigh D. O., Interphase Phenomena in Solid Electrochemical Cells, *Fast-Ion Transport in Solids*, ed. van Gool W. (North-Holland, Amsterdam, 1972) p.479
79. Xu Y., Chen C. J., Xu R., and MacKenzie J. D., Ferroelec. **108**, 1653 (1990)

Chapter 3

80. O'Dwyer J. J., *Theory of Dielectric Breakdown in Solids* (Clarendon Press, Oxford, 1964); Wolters D. R. and Zegers-van Duijnhoven A. T. A., J. Vac. Sci. Technol. **A5**, 1563 (1987); Coelho R., *Physics of Dielectrics* (Elsevier, Lausanne, 1979); Waser R. and Smyth D. M., *Ferroelectric Thin Films*, eds. Paz de Araujo C. A., Scott J. F., and Taylor G. W. (Gordon & Breach, New York, 1996) p.47; Williams R., Phys. Rev. **125**, 850 (1962)
81. No K., et al., MRS Proc. **433**, 9 (1996); Seifert S.,, Loebmann P., and Sporn D., J. Phys. IV France **8**, 61 (1998)

82. Vlasenko N. A., Opt. i Spekt. **18**, 260 (1965)
83. Goetzberger A., McDonald B., Haitz R. H., and Scarlett R. M., J. Appl. Phys. **34**, 1591 (1980)
84. Pearsall T. P., J. Appl. Phys. **34**, 1591 (1980)
85. Waser R. and Klee M., Integ. Ferroelec. **2**, 288 (1990)
86. Von Hippel A., J. Appl. Phys. **8**, 815 (1937)
87. Von Hippel A., Ergeb. exakt Naturwiss. **14**, 118 (1935)
88. Von Hippel A., Z. Phys. **98**, 580 (1936)
89. Dekker A. J., Phys. Rev. **94**, 1179 (1954)
90. Dakin T. W. and Berg D., *Progress in Dielectrics*, eds. J. B. Birks and J. Hart (Academic Press, New York, 1962) Vol. **4**, p.165
91. Sharbaugh H. and Watson P. K., *Progress in Dielectrics*, eds. J. B. Birks and J. Hart (Academic Press, New York, 1962) Vol. **4**, p.226
92. Scott J. F., Azuma M., et al., Integ. Ferroelec. **4**, 61 (1994)
93. Nowotny J. and Sloma N., *Surface and Near-Surface Chemistry of Oxide Ceramic Materials*, eds. J. Nowotny and L. C. Dufour (Elsevier, Amsterdam, 1988)
94. de Boer J. H., *Electron Emission and Absorption Phenomena* (Cambridge University Press, London, 1935)
95. Zafar S., Hradsky B., et al., J. Appl. Phys. (1999 in press)
96. Scott J. F., Melnick B. M., McMillan L. D., and Paz de Araujo C. A., Integ. Ferroelec. **3**, 225 (1993)
97. Gerson R. and Marshall T. C., J. Appl. Phys. **30**, 1650 (1959)
98. Matsubara S. et al., MRS Proc. **200**, 243 (1990)
99. Matsubara S. et al., MRS Proc. **243**, 281 (1992)
100. Carrano J., et al., IEDM Conf. Proc. (IEEE, New York, 1989) p.225
101. Sudhama C. et al., MRS Proc. **200**, 331 (1990)
102. Arita K., Proc. ISAF, (IEEE, New York, 1996)
103. Scott J. F., *Science and Technology of Electroceramic Thin Films*, eds. O. Auciello and R. Waser, NATO ASI Series **284** (Kluwer, Dordrecht, 1995) p.249
104. Scott J. F., Ross F. M., et al., MRS Bull. **21**, 33 (1996); Ferroelectrics **201**, 43 (1997)
105. Feng Duan, private communication (1991)
106. Srolovitz D. J. and Scott J. F., Phys. Rev. **B34**, 1815 (1986)
107. Plumlee R., Sandia Laboratories Report SC-RR-67-730 (1967)
108. Bursill L., private communication (1996)
109. De Veirman A., E. M., et al. EMIF-1, Nijmegen (July 4, 1995); Ferroelectrics **186**, 51 (1996)
110. Lipeles R. A., Morgan B. A., and Leung M. S., Integ. Ferroelec. **5**, 197 (1994)
111. Raleigh D. O., *Fast-Ion Transport in Solids*, ed. W. van Gool (North-Holland, Amsterdam, 1972) p.479
112. Duiker H. M.,, Ph.D. thesis, Univ. Colorado (1990)
113. Duiker H. M. and Beale P. D., Phys. Rev. **B41**, 490 (1990)
114. Duiker H. M. et al., J. Appl. Phys. **68**, 5783 (1990)
115. Freund F. et al., Le Vide: Science, Technique et Applications **275**, Suppl. 538 (1995) [Proc. 2nd Int. Conf. Space Charge in Solid Dielectrics, Antibes, 1995]
116. Scott J. F. et al., J. Appl. Phys. **64**, 787 (1988)
117. Eaton S., private communication (1992)
118. Seitz F., Phys. Rev. **76**, 1328 (1949)
119. Waser R. and Klee M., Integ. Ferroelec. **2**, 288 (1992)
120. Forlani F. and Minnaja N., Phys. Stat. Sol. **4**, 311 (1964)
121. Klein N. and Gafni H., IEEE Trans. Electron. Dev. **13**, 281 (1966)
122. Klein N. and Levanon N., J. Appl. Phys. **38**, 3721 (1967)
123. Klein N. and Levanon N., J. Electrochem. Soc. **116**, 963 (1969)

124. Harrop P. J. and Campbell D. S., *Handbook of Thin Film Technology*, eds. Maissel L. I. and Glang R. (McGraw-Hill, New York, 1970) p.16
125. Zuleeg R. and Miller R. S., Sol. St. Electron. **7**, 575 (1964)
126. Tagantsev A. K., Kholkin A. L., Colla E. L., Brooks K. G., and Setter N., Integ. Ferroelec. **10**, 189 (1995)
127. Bernacki S. E., MRS Proc. **243**, 135 (1992)
128. Scott J. F., Galt D., et al., Integ. Ferroelec. **6**, 189 (1995)
129. Watanabe K., private communication
130. Scott J. F., Integ. Ferroelec. **232**, 25 (1999)

Chapter 4

131. Sze S. M., *Physics of Semiconductor Devices*, 2nd ed. (Wiley, New York, 1981) p.252
132. Scott J. F., Azuma M., et al., Integ. Ferroelec. **4**, 61 (1994)
133. Lee J.-J., Alluri P. and Dey S. K., Appl. Phys. Lett. **65**, (1994)
134. Dietz G. W., PhD thesis, Tech. Univ. Aachen (1996)
135. Joshi V., DaCruz C. P., Cuchiaro J. D., Araujo C. A., and Zuleeg R., Integ. Ferroelec. **14**, 133 (1997); Diemer G., Physica **26**, 889 (1960); Waser R. and Smyth D. M., *Ferroelectric Thin Films*, eds. Paz de Araujo C. A. et al. (Gordon & Breach, New York, 1996), p.47
136. Smyth D. M., Prog. Sol. St. Chem. **15**, 145 (1984); Chan N.-H., Sharma R. K., and Smyth D. M., J. Am. Ceram. Soc. **64**, 556 (1981); J. Electrochem. Soc. **123**, 1584 (1976)
137. Wouters D. J., Wilems G. J., and Maes H. E., Microelectron. Eng. **29**, 249 (1995); this erroneous band structure was subsequently used by Stolichnov I. and Tagantsev A. K., J. Appl. Phys. **84**, 3216 (1998); Dixit A. V., Rajopadhye N. R., and Bhoraskar S. V., J. Mater. Sci. 21, 2798 (1986)
138. Scott J. F., Ferroelec. **183**, 51 (1996); see also the earlier work by Merz W. and Anderson J. R., Bell Labs. Record **33**, 335 (1955) and by Campbell D. S., Phil. Mag. **79**, 1157 (1962). For Ti^{3+} studies see Scott J. F. and Ross F. M., Ferroelec. **201**, 43 (1997) and Bardi U., Tamura K., Owari M., and Nihei Y., Appl. Surf. Sci. **2**, 352 (1988)
139. Mark P. and Hartman T. E., J. Appl. Phys. 39, 2163 (1968); Li P. and Lu T. M., Phys. Rev. **B43**, 14261 (1991)
140. Robblee L. S., US Pat. No. 4,677,989 (1987) and 4,717,581 (1987)
141. Robblee L. S., et al., MRS Proc. **55**, 303 (1986)
142. Robblee L. S. and Cogan S. F., *Metals for Medical Electrodes*, Enc. Mat. Sci. & Eng., Suppl. Vol. **1**, ed. R. W. Cahn (Pergamon, Oxford, 1986) p.276
143. Williams D. F., Ann. Rev. Mat. Sci. **6**, 237 (1976)
144. Robblee L. S. and Rose T. L., *Neural Prostheses: Fundamental Studies*, ed. W. Agnew and D. McCreery (Prentice-Hall, New York, 1989) p.117
145. Uhlig H. H. and Reeve R. W., *Corrosion and Corrosion Control* (Wiley, New York, 1985)
146. Brummer S. B., Robblee L. S., and Hambrecht F. T., Ann. NY Acad. Sci. **405**, 159 (1983)
147. Brummer S. B. and Turner M. J., IEEE Trans. Biomed. Eng. **24**, 59 (1977)
148. Kingon A., private communication; note that the oxidation of Pt in applied electric fields is very different from that at $V = 0$, which occurs only at elevated temperatures: *Smithells Metals Reference Book*, 6th ed., Brandes E. A., ed. (Butterworths, London, 1983) Chap.13 Diffusion in Metals

149. Noguchi T. et al., FMA-13, Kyoto (1996), pp.27 and 78; Scott J. F., Ferroelec. Rev. **1**, 1 (1998)
150. Smith R. W. and Rose A., Phys. Rev. **97**, 1531 (1955); Ruppel W., Helv. Phys. Acta **31**, 311 (1958)
151. Hamann C., Burghardt H., and Frauenheim T., *Electrical Conduction Mechanisms in Solids* (VEB Publishing Co., Berlin, 1988); Lampert M. A., Many A., and Mark P., Phys. Rev. **135**, A1444 (1964)
152. Scott J. F., Melnick B. M. et al., Integ. Ferroelec. **4**, 85 (1994); **3**, 225 (1993)
153. Melnick B. M., Scott J. F., Araujo C. A., and McMillan L. D., Ferroelec. **135**, 163 (1992)
154. Peng C. J., Hu H., Krupanidhi S. B.. Appl. Phys. Lett. **63**, 1038 (1993)
155. Mayer J. W., Baron R., and Marsh O. J., Phys. Rev. **137**, A286 (1965)
156. Rose A., Phys. Rev. **97**, 1538 (1955)
157. Lampert M. A. and Mark P., *Charge Injection in Solids* (Academic Press, New York, 1970) p.73, p.276
158. Scott J. F., Watanabe K., and Hartmann A. J., Integ. Ferroelec. **21**, 241 (1998)
159. Lampert M. A. and Mark P., *Charge Injection in Solids* (Academic Press, New York, 1970) p.261
160. Waser R., *Science and Technology of Electroceramic Thin Films* (Kluwer, Dordrecht, 1995) p.223
161. Brennan C., Integ. Ferroelec. **8**, 335 (1995)
162. Brennan C., Integ. Ferroelec. **8**, 93 (1995)
163. Scott J. F., et al., Integ. Ferroelec. **4**, 85 (1994)
164. Roy D., Peng C. J., and Krupanidhi S. B., Appl. Phys. Lett. **60**, 2478 (1992)
165. Many A. and Raklavy G., Phys. Rev. **126**, 1980 (1962)
166. Zafar S. et al., Appl. Phys. Lett. **73**, 175 (1998)
167. Chen H. D. et al., Integ. Ferroelec. **15**, 89 (1997); Kroger F. A., Diemer G., and Klasens H. A., Phys. Rev. **103**, 279 (1956)
168. Tredgold R. H., *Space Charge Conduction in Solids* (Elsevier, Amsterdam, 1966); reviews of double injection are given by Ivey H. F., J. Electrochem. Soc. **180**, 590 (1961) and by Henisch H. K., *Electroluminescence* (Macmillan, New York, 1962)
169. Rose A., Phys. Rev. **97**, 1537 (1955)
170. Wu Z. and Sayer M., Proc. ISAF (IEEE, New York, 1992) p.244
171. Waser R., Baiatu T. and Hardtl K.-H., J. Am. Ceram. Soc. **73**, 1645 (1990); **73**, 1654 (1990); **73**, 1663 (1990)
172. Mihara T., et al., Integ. Ferroelec. **1**, 269 (1992); Nikkei Electronics **581**, 94 (1993)
173. Wang D. Y. and Umeya K., J. Am. Ceram. Soc. **73**, 1574 (1990)
174. Dey S. K. and Zuleeg R., Ferroelec. **109**, 1643 (1990); Dey S. K., Alluri P., and Lee J.-J., Integ. Ferroelec. **7**, 341 (1993)
175. Klimov V. V., Ferroelec. **17**, 465 (1978)
176. Jo W. et al., MRS Proc. **361**, 33 (1994)
177. Abe K., Komatsu S., Yanase N., Sano K., and Kawakubo T., Jpn. J. Appl. Phys. **36**, 5846 (1997)
178. Abe K. et al., IEICE **81**, 505 (1998)
179. Kreyina G. S., Selivanov L. N., and Shumskaia T. I., Radio Eng. Elec. Phys. **5**, 8 (1960); **5**, 219 (1960); **7**, 166 (1962); **7**, 1949 (1962)
180. Hickmott T. W., J. Appl. Phys. **33**, 2669 (1962); **34**, 1569 (1963); **35**, 2118 (1964); **35**, 1885 (1965)
181. Lewowski T., Sendecki S., and Sujak B., Acta Phys. Polon. **28** (1965)
182. Uzan R., Roger A., and Cachard A., Vide **137**, 38 (1967)

183. Simmons J. G. and Verderber R. R., Proc. Roy. Soc. (London) **301**, 77 (1967); Radio Elec. Eng. **34**, 81 (1966)
184. Verderber R. R., Simmons J. G., and Eales B., Phil. Mag. **16**, 1049 (1967)
185. Klein N. and Levanon N., J. Appl. Phys. **38**, 3721 (1967); Pratt R. G. and Ridley B. K., Proc. Phys. Soc. (London) **81**, 996 (1963); **85**, 293 (1965); J. Phys. Chem. Sol. **26**, 11 (1965); **26**, 21 (1965); Phys. Lett. **4**, 300 (1963)
186. Kostic P., Milosevic O., Uskokovic D., and Ristic M. M., Physica **B150**, 175 (1988)
187. Ridley B. K., Proc. Phys. Soc. (London) **82**, 954 (1963); Barnett A. M. and Milnes A. G., J. Appl. Phys. **37**, 4215 (1966)
188. Nicoll F. M., RCA Rev. **19**, 17 (1958); Bube R. H., *Physics and Chemistry of II–VI Semiconductors*, eds. Aven M. and Prener J. S. (North-Holland, Amsterdam, 1967) p.657

Chapter 5

189. Brennan C. J., Integ. Ferroelec. **2**, 73 (1992) gives a good review
190. Kalkur T. S. et al., MRS Proc. **230**, 315 (1992)
191. Mihara T. et al., Integ. Ferroelec. **1**, 269 (1992)
192. Hwang C. S. et al., Mater. Sci. & Eng. **B56**, 178 (1998)
193. Hwang C.-S. et al., J. Appl. Phys. **83**, 3703 (1998); **85**, 287 (1999)
194. Scott J. F., Azuma M., et al., Proc. ISAF (IEEE, New York 1992) p.356
195. Sayer M., Mansingh A., Arora A. K., and Lo A., Integ. Ferroelec. **1**, 129 (1992) treat this problem correctly
196. Evans J., IEEE ISAF (Urbana, IL, 7 June 1990)
197. Miller S. L. et al., J. Appl. Phys. **68**, 6463 (1990)
198. Weisberg L. R., J. Appl. Phys. **33**, 1817 (1962)
199. Nelson D. F., Johnson L. F., and Gershenzon, Bull. Am. Phys. Soc. **9**, 236 (1964)
200. Bube R. H., Photoelectronic Analysis, *Semiconductors and Semimetals*, eds. Willardson R. K. and Beer A. C. (Academic Press, New York, 1967) p.461
201. Outzourhit A., et al., Integ. Ferroelec. **8**, 227 (1995)
202. Dietz G. W., PhD thesis, Tech. Univ. Aachen (1996); Zafar S. et al., Appl. Phys. Lett. **73**, 1 (1998)
203. Dietz G. W. et al., J. Appl. Phys. **82**, 2359 (1997)
204. Sze S. M., *Physics of Semiconductor Devices*, 2nd ed. (Wiley, New York, 1981)
205. Lampert M. A. and Mark P., *Current Injection in Solids* (Academic Press, New York, 1975)
206. Mihara T., Nikkei Electronics **581**, 94 (1993)
207. Copel M. et al., Appl. Phys. Lett. **70**, 3227 (1997); Moazammi R., Hu C., and Shepherd W. H., IEEE Trans. Electron. Dev. **39**, 2044 (1992)
208. Blom P. W. M., Wolff R. M., Cilessen J. F. M., and Krijn M. P. C. M., Phys. Rev. Lett. **73**, 2107 (1994)
209. Scott J. F., Proc. AMF-2 (Singapore, Dec. 1998), Ferroelectrics (in press 1999); Proc. SSDM (Jpn. Solid State Device Meeting), Hiroshima, Sept. 1998 [Jpn. J. Appl. Phys. **38**, 2272 (1999)]
210. Jo W., Kim D. C., Lee K. H., and Kim K. Y., J. Korean Phys. Soc. **34**, 61 (1999); **32**, S1442 (1998)
211. Dey S. K., Proc. AMF-2 (Singapore Dec. 1998), Ferroelectrics (in press 1999)

212. Simmons J. G., J. Phys. Chem. Sol. **32**, 1987 (1971); **32**, 2581 (1971); Phys. Rev. Lett. **15**, 967 (1965); Frank R. I. and Simmons J. G., J. Appl. Phys. **38**, 832 (1967); Zafar et al., Appl. Phys. Lett. **73**, 1 (1998)
213. Frank G. and Simmons J. G., J. Appl. Phys. **38**, 832 (1967)
214. Geppert D. V., J. Appl. Phys. **33**, 2993 (1962); Meyerhofer A., unpublished (reproduced in [205], p.190); Kahn A. H. and Leyendecker A. J., Phys. Rev. **135A**, 1321 (1964)
215. Many A. and Raklavy G., Phys. Rev. **126**, 1980 (1962)
216. Adirovich E. I., Sov. Phys. Sol. St. **2**, 1282 (1960) [Fiz. Tverd. Tela **2**, 1410 (1960)]
217. Stoeckmann F., Phys. St. Sol. **3**, 221 (1963)
218. Chai F. K., Brews J. R., Schrimf R. D., and Birnie D. P., J. Appl. Phys. **78**, 4766 (1995)
219. Chai F. K. et al., J. Appl. Phys. **82**, 2517 (1997)
220. Scott J. F., Azuma M., et al., Integ. Ferroelec. **4**, 61 (1994); Wang Y.-P. and Tseng T.-Y., J. Appl. Phys. **81**, 6762 (1997)
221. Brillson L. J., Surf. Sci. **299**, 909 (1994)
222. Kurtin S., McGill T. C., and Mead C. A., Phys. Rev. Lett. **12**, 390 (1964)
223. Bardeen J., Phys. Rev. **71**, 717 (1947)
224. Mueller R. S. and Kamins T. I., *Device Electronics for Integrated Circuits* (Wiley, New York, 1977) p.93
225. Robertson J. and Chen C. W., MRS Proc. (Dec 1998 meeting) (1999)
226. Schluter M., Phys. Rev. **B17**, 5044 (1978); Thin Solid Films **93**, 3 (1982)
227. Wouters D. J., Wilems G. J., and Maes H. E., Proc. EMF-8, Microelectron. Eng. **29**, 249 (1995)
228. Dey S. K., Alluri P., Lee J.-J., and Zuleeg R., Integ. Ferroelec. **8**, 715 (1995); Dey S. K. and Zuleeg R., Ferroelec. **109**, 1643 (1990)
229. Klissurska R. D., PhD thesis, EPFL, Lausanne (1998); Klissurska R. D., Brooks K. G., and Setter N., Proc. ISAF (Lausanne, Sep. 1998)
230. Scott J. F., Proc. Japan SSDM (Hiroshima, Sept. 1998), Jpn. J. Appl. Phys. (in press 1999)
231. Hwang C. S., et al., J. Appl. Phys. (1999 in press)
232. Lampert M. A., Rep. Prog. Phys. **27**, 329 (1964); Lampert M. A. and Mark P., *Current Injection in Solids* (Academic Press, New York, 1975) p.261
233. Scott J. F. and Hartmann A. J., Proc. ISAF (Lausanne, Sept. 1998), Ferroelectrics (in press 1999)
234. Dacey G. C., Phys. Rev. **90**, 759 (1953)
235. Colla E., Tagantsev A. K., et al., ISIF 1999 (Integ. Ferroelec., in press)
236. Joshi V., DaCruz C. P., Cuchiaro J. D., Araujo C. A., and Zuleeg R., Integ. Ferroelec. **14**, 133 (1997); Lee S. C. et al., Integ. Ferroelec. **4**, 31 (1994); Carrico A. S., Paz de Araujo C. A., Mihara T., and Watanabe H., Integ. Ferroelec. **13**, 247 (1996)
237. Scott J. F., IEICE Trans. Electron. **81**, 477 (1998)
238. Scott J. F., Ferroelec. Rev. **1**, 1 (1998)
239. Jo W., LG Semiconductor (private communication, March 1999); Tanaka M. and Ami M., J. Am. Ceram. Soc. **81**, 2513 (1998); Zafar S., et al., Appl. Phys. Lett. **73**, 1 (1998); Glassford K. M. and Chelikowski J. R., Phys. Rev. **B47**, 1732 (1993); **B49**, 7107 (1994); Koetz R. and Stucki S., Electrochim. Acta 31, 1311 (1986)
240. Seong N.-J. and Yoon S.-G., J. Phys. IV France **8**, 255 (1998)

241. Zafar S. et al., Appl. Phys. Lett. **73**, 1 (1998); Choi G. M., Tuller H. E., and Goldschmidt D., Phys. Rev. **B34**, 6972 (1986); Many A. and Rakavy G., Phys. Rev. **26** 1989 (1962). For ionic space charge limited currents see Hofstein S. R., Appl. Phys. Lett. **10**, 291 (1967); **11**, 95 (1967) and Lemke H. and Mueller G. O., Phys. Stat. Sol. **24**, 127 (1967)
242. Appel J., Polarons, *Solid State Physics*, eds. seitz F., Turnbull D., and Ehrenreich H. (Academic, New York, 1968) **21**, 193; Chen X. and Kingon A. I., Proc. ISAF (IEEE, New York, 1962) ed. S. Liu, p.75
243. Koops C. S., Phys. Rev. **83**, 121 (1951)
244. Heywang W., Z. Angew. Phys. **16**, 1 (1963)
245. Gersthsen P. and Hardtl K.-H., Z. Naturf. **18a**, 423 (1963)
246. Jonker G. H., Sol. St. Electron. 7, 865 (1964)
247. Frederikse H. P. R., Thurber W. R., and Hosler W. R., Phys. Rev. **134**, A442 (1964); Phys. Rev. **161**, 822 (1967)
248. Tufte O. and Chapman P. W., Phys. Rev. **155**, 796 (1967)
249. Smith R. W. and Rose A., Phys. Rev. **97**, 1531 (1955)

Chapter 6

250. Landau L. D., Phys. Z. Sowjetunion **11**, 26 (1937)
251. Bune A. V., Fridkin V. M., Ducharme S., et al., Nature **391**, 874 (1998)
252. Kay H. F. and Dunn J. W., Phil. Mag. **7**, 2027 (1962)
253. Scott J. F. et al., Proc. ISAF (IEEE, New York, 1986) p.569
254. Scott J. F., Ferroelectrics Rev. **1**, 1 (1998)
255. Bellur K. R., Al-Shareef H. N., and Rou S. H., Proc. ISAF (IEEE, New York, 1992) p.448
256. Sakashita Y., Ono T., Segawa H, Tominaga K., and Okada M., J. Appl. Phys. **73**, 7857 (1993)
257. Shur V., et al., Integ. Ferroelec. **2**, 51 (1992) and private communication cited in [254]
258. Janovec V., Phys. Lett. **99A**, 549 (1983)
259. Democritov S. O., Kreines N. M., et al., JETP Lett. **48**, 294 (1988)
260. Seul M. and Andelman D., Science **267**, 476 (1995)
261. Muratov C. B. and Osipov V., Phys. Rev. **E53**, 3101 (1996)
262. Stadler H. L. and Zachminidis P. J., J. Appl. Phys. **29**, 1285 (1958); **33**, 3487 (1962); **34**, 3255 (1963)
263. Duiker H. M. et al., J. Appl. Phys. **68**, 5783 (1990)
264. Ishibashi Y., Integ. Ferroelec. 3, 351 (1993); Jpn. J. Appl. Phys. **24**, Suppl. 3, 126 (1986)
265. Dennis J., Integ. Ferroelec. **3**, 1 (1993)
266. Scott J. F. et al., J. Appl. Phys. **64**, 787 (1988)
267. Orihara H. and Ishibashi Y., J. Phys. Soc. Jpn. **61**, 1919 (1992); **61**, 4650 (1992)
268. Lines M. and Glass A. M., *Principles and Applications of Ferroelectrics and Related Materials* (Clarendon, Oxford, 1977)
269. Dalton N. W., Jacobs J. T. and Silverman B. D., Phys. Rev. **133A**, 1034 (1964)
270. Avrami M., J. Chem. Phys. **7**, 1103 (1940); **8**, 212 (1940); **9**, 177 (1941)
271. Hase T. and Shiosaki T., Jpn. J. Appl. Phys. **30**, 2159 (1991)
272. Sekimoto K., Physica **A135**, 328 (1986); **A137**, 96 (1986)
273. Ishibashi Y., Integ. Ferroelec. **9**, 57 (1995)

274. Franke K., Besold J., Haessler W., and Seegebarth C., , Surf. Sci. **302**, L283 (1994)

Chapter 7

275. Scott J. F. and Pouligny B., J. Appl. Phys. **64**, 1547 (1988)
276. Al-Shareef H. N., Dimos D., Warren W. L., and Tuttle B. A., Integ. Ferroelec. **15**, 53 (1997)
277. Dimos D. et al., J. Appl. Phys. **76**, 4305 (1994); Integ. Ferroelec. **2**, 477 (1992)
278. Warren W. L., Tuttle B. A., and Dimos D., Appl. Phys. Lett. **67**, 1426 (1995); Warren W. L. et al., J. Appl. Phys. **77**, 6695 (1995); MRS Bull. **21**, 40 (1996)
279. de Wette F., NATO Conf. Electron–Phonon Interactions (Erice, Sicily, June 1997) unpublished manuscript
280. Colla E. L., Taylor D. V., Tagantsev A. K., and Setter N., Appl. Phys. Lett. (in press); Lee H. S. and Lee K. B., J. Phys. IV France **8**, 209 (1998)
281. Colla E. L., Tagantsev A. K., Tasylor D. V., and Kholkin A. L., Integ. Ferroelec. **22**, 1045 (1998); J. Korean Phys. Soc. **32**, S1353 (1998)
282. Gruverman A., Auciello O., and Tokumoto H., Appl. Phys. Lett. **69**, 3191 (1996); Integ. Ferroelec. **19**, 49 (1998)
283. Melnick B. M., Scott J. F., Paz de Araujo C. A., and McMillan L. D., Ferroelec. **135**, 163 (1992)
284. Zafar S., Hradsky B., Gentile D., Chu P., Jones R. E., and Gillespie S., J. Appl. Phys. (in press 1999)
285. Nowotny J. and Sloma N., *Surface and Near-Surface Chemistry of Oxide Ceramic Materials*, eds. Nowotny J. and Dufour L. C. (Elsevier, Amsterdam, 1988)
286. Waser R., Baiatu T., and Hardtl K., J. Am. Ceram. Soc. **73**, 1645 (1990); **73**, 1654 (1990); **73**, 1663 (1990)
287. Arlt G. and Neumann H., Ferroelectrics **87**, 109 (1988); Arlt G. and Robels U., Integ. Ferroelec. **3**, 343 (1993)
288. Dawber M., B. S. Honours Thesis, Univ. New South Wales (1999); Dawber M. and Scott J. F., Appl. Phys. Lett. **76**, 1060 (2000)
289. Brennan C., Ferroelectrics **150**, 199 (1993)
290. Yoo I. K. and Desu S. B., Integ. Ferroelec. 3, 365 (1993); Desu S. B., Phys. Stat. Sol. A151, 467 (1995)
291. Dey S. K., Ferroelectrics (in press 1999)
292. Mihara T., Watanabe H., and Paz de Araujo C. A., Jpn. J. Appl. Phys. **33**, 3996 (1994)
293. Scott J. F., Integ. Ferroelec. **1**, 1 (1998)
294. Zafar S. et al., Appl. Phys. Lett. **73**, 1 (1998)
295. Smyth D., *Ferroelectric Thin Films* (Gordon & Breach, New York, 1996) eds. C. Araujo, J. F. Scott, and G. W. Taylor
296. Colla E., Taylor D. V., Tagantsev A. K., Setter N., Appl. Phys. Lett. **72**, 2478 (1998); Colla E. L., Hong S., Taylor D. V., Tagantsev A. K., and Setter N., Appl. Phys. Lett. **72**, 2763 (1998); Colla E. L., et al., Microelec. Eng. **29**, 145 (1995)
297. Moritomo A., Yamanaka Y., and Shimizu T., Jpn. J. Appl. Phys. **34**, 4108 (1995); Paton E., Brazter M., Mansour S., and Bement A., Integ. Ferroelec. **18**, 29 (1997)
298. Tagantsev A. K., Landivar M., Colla E., and Setter N., J. Appl. Phys. **78**, 2623 (1995)

299. Larsen P. K., Dormans G. J. M., Taylor D. J., and van Veldhoven P. J., J. Appl. Phys. **76**, 2405 (1994)
300. Fouskova A., J. Phys. Soc. Jpn. **27**, 1699 (1969)
301. Majunder S. B., Mohapatra Y. N., and Agrawal D. C., Appl. Phys. Lett. **70**, 138 (1997); Taylor D. V., Damjanovic D., Colla E., and Setter N., Ferroelectrics **225**, 91 (1999)
302. Du X. and I-Wei Chen, J. Appl. Phys. **83**, 7789 (1998)
303. Grenier J. C., Durier J., Pouchard M., and Hagenmuller P., Mat. Res. Bull. **11**, 1219 (1976)
304. Grenier J. C., Schiffmacher G., Caro P., Pouchard M., and Hagenmuller P., J. Sol. St. Chem. **20**, 365 (1977)
305. Grenier J. C., Pouchard M., and Hagenmuller P., Structure and Bonding **47**, 1 (1981)
306. Beccero A. I., McCammon C., Lagenhorst F., Seifert F., and Angel R., Phase Transitions **69**, 133 (1999)
307. Khachaturyan A. G., *Theory of Structural Transformations in Solids* (Wiley, New York, 1983), pp.511–523
308. Arlt G. and Neumann H., Ferroelectrics **87**, 109 (1988); Arlt G. and Roebels U., Integ. Ferroelec. **3**, 343 (1993); Roebels U., Calderwood J. H., and Arlt G., J. Appl. Phys. **77**, 4002 (1995)
309. Brennan C., Integ. Ferroelec. **8**, 93 (1995); **8**, 335 (1995)
310. Yoo I. K. and Desu S. B., Phys. Stat. Sol. **A133**, 565 (1992)
311. Dawber M. and Scott J. F., Appl. Phys. Lett. (in press, 1999)
312. Colla E. L., Taylor D. V., Tagantsev A. K., and Setter N., Appl. Phys. Lett. **72**, 2478 (1998)
313. Paton E., Brazier M., Mansour S., and Bement A., Integ. Ferroelec. **18**, 29 (1997)
314. Mihara T., Watanabe H., and Paz de Araujo C. A., Jpn. J. Appl. Phys. **33**, 3996 (1994)
315. Zhang G. B. and Smyth D. M., Sol. St. Ionics **82**, 161 (1995)
316. Scott J. F., Azuma M., Fujii E., Otsuki T., Kano G., Scott M. C., Paz de Araujo C. A., McMillan L. D., and Roberts T., Proc. ISAF, Greenville, SC; IEEE (catalog 92CH3080-9), eds. Liu M. et al., (IEEE, New York, 1992), pp.356–9
317. Stolichnov I., Tagantsev A. K., Colla E. L., and Setter N., Appl. Phys. Lett. **73**, (1998)
318. Steinsvik S., Bugge R., Gjonnes J., Tafto J., and Nordby T., J. Phys. Chem. Sol. **58**, 969 (1997)
319. Wang Y. G., Steinsvik S., Haier R., and Nordby T., J. Mater. Sci. Lett. **14**, 1027 (1995)
320. Sutija D., Nordby T., Osborg P. A., and Kofstad P., Electrochem. Soc. Proc. **93–4**, 552 (1993)
321. Eaton S., cited by Scott J. F., Ferroelectrics Rev. **1**, 1 (1998)
322. Guettler B., Bismayer U., Groves P., and Salje E., Semicond. Sci. Technol. **10**, 245 (1995)
323. Paz de Araujo C. A., Cuchiaro J. D., McMillan L. D., Scott M. C., and Scott J. F., Nature **374**, 627 (1995)
324. Uskov M. P. and Khachaturyan A. G., Sov. Phys. Crystall. **13**, 910 (1969)
325. van Landuyt J., Gevers R., and Amelinckx S., Phys. Stat. Sol. **13**, 467 (1966)

Chapter 8

326. Ishibashi Y. and Orihara H., Integ. Ferroelec. **9**, 57 (1995)
327. Avrami M., J. Chem. Phys. **9**, 177 (1941)
328. Scott J. F., Ross F. M., Paz de Araujo C. A., Scott M. C., and Huffman M., MRS Bull. **21**, 33 (1996)
329. DeVilbis A. et al., ISIF 1996 (Tempe, AZ), unpublished
330. Ramesh R. (private communication)
331. Waser R. et al., AMF-2, Singapore, Dec 8, 1998 (Ferroelectrics, in press)
332. Zafar S. et al., Appl. Phys. Lett. **72**, 2820 (1998)
333. Scott J. F., Galt D., et al., Integ. Ferroelec. **6**, 189 (1995)

Chapter 9

334. Wersing W., Siemens Corp., Eur. Conf. Polar Devices (Bled, Slovenia, 1996); Iembo A. et al., Integ. Ferroelec. **18**, 397 (1997)
335. Zhou W., J. Sol. St. Chem. **101**, 1 (1992)
336. Rodriguez M. A., Boyle T. J., et al., Proc. ISIF 1996
337. Hartmann A. J., et al., Ferroelec. Lett. **23**, 75 (1997)
338. Rae D., Thompson J. G., and Withers R. L., Acta Cryst. **B48**, 418 (1992); **B46**, 474 (1990)
339. Subbarao E. C., J. Chem. Phys. **34**, 695 (1961); Phys. Rev. **122**, 804 (1961)
340. Atsuki T., Soyama N., Yonezawa T., and Ogi K., Jpn. J. Appl. Phys. **34**, 5096 (1995)
341. Noguchi T., Hase T., and Miyasaka Y., Jpn. J. Appl. Phys. **35**, 4900 (1996)
342. Watanabe K., Tanaka M., Sumitomo E., Katori K., Yagi H., and Scott J. F., Appl. Phys. Lett. **73**, 126 (1998); pure SBN has also been studied by Yi J. H. et al., J. Phys. IV France **8**, 225 (1998)
343. Kapitza P., Proc. Roy. Soc. (London) **A119**, 428 (1928)
344. Bussem W., Gross F., and Hermann K., Ann. Phys. **64**, 537 (1930)
345. Suhrmann R. and Schnackenberg H., Z. Phys. **119**, 287 (1942)
346. Van Itterbeck A. and de Bock A., Ann. Phys. **11**, 635 (1945)
347. Sondheimer E. H., Adv, Phys. **1**, 1 (1952)
348. Pashley D. W., Adv. Phys. **5**, 173 (1956)
349. Heine V., Proc. Phys. Soc. **A69**, 513 (1956)
350. Colombani A. and Huet P., *Structure and Properties of Thin Films*, eds. Neugebauer C. A., Newkirk J. B., and Vermilyea D. A. (Wiley, New York, 1959) p.253
351. Buckel W., Ibid., p.53

Chapter 10

352. Tachiki M. et al., Jpn. J. Appl. Phys. **38**, L179 (1996)
353. Reaney I. M., Euroceram V (Aveiro, Sept. 1996) Vol. 1, p.443; the superlattice diffraction lines for the titanates are also visible in Scott J. F. and Ross F. M., Proc. ECAPD (Bled, Slovenia, August 1996), Ferroelectrics **201**, 43 (1997)

354. Poignant F., Trolliard G., and Abelard P., Proc. Electroceramics IV, ed. Waser R. (Augustinus Buchhandlung Verlag, Aachen, 1994) p.639; Yu Z., Chen A. et al., Proc. Eur. Conf. Ceramics (Aveiro, Portugal, Sept. 1996) p.491; Tabata H., Hamada M., and Kawai T., MRS Proc. **401**, 73 (1996)
355. Hase T., Noguchi T., and Miyasaka Y., Integ. Ferroelec. **16**, 29 (1997); Zafar S. et al., J. Appl. Phys. **82**, 4469 (1997)
356. Ami T. et al., MRS Proc. **415**, 195 (1995)
357. Ashida H. et al., Integ. Ferroelec. **21**, 97 (1998)
358. Takatani S., Kushida-Abdelghafar K., and Miki H., Jpn. J. Appl. Phys. **36**, L453 (1997)
359. Fujisaki Y., Kushida-Abdelghafar K., Shimamoto Y., and Miki H., J. Appl. Phys. **82**, 341 (1997)
360. Hartmann A. J., Scott J. F., et al., J. Korean Phys. Soc. **32**, S1329 (1998); Scott J. F., Ferroelec. Rev. **1**, 1 (1998)
361. Jones R. E., Jr., et al., Thin Solid Films **270**, 584 (1995)
362. Miki H., Kushida-Abdelghafar K., Torii K., and Fujisaki Y., Jpn. J. Appl. Phys. **36**, 1132 (1997)
363. Kushida-Abdelghafar K., Miki H., Torii K., and Fujisaki Y. J. Appl. Phys. **69**, 3188 (1996)
364. Han J.-P. and Ma T. P., Integ. Ferroelec. **22**, 733 (1998)
365. Warren W. L. et al., J. Appl. Phys. **67**, 1426 (1995)
366. Dimos D. et al., J. Appl. Phys. **76**, 4305 (1994)
367. Warren et al., Integ. Ferroelec. **16**, 77 (1997)
368. Al-Shareef H. N., Dimos D., Warren W. L., and Tuttle B. A., Integ. Ferroelec. **15**, 53 (1997)
369. Troeger G., McDonnell-Douglas Corp., private communication cited in Scott J. F., et al., J. Appl. Phys. **70**, 382 (1991)
370. Stannard W. B. et al., Microelectron. Eng. **29**, 193 (1995); Integ. Ferroelec. **9**, 245 (1995)
371. Watanabe K., Scott J. F. et al. Integ. Ferroelec. **22**, 241 (1998)
372. Hartmann A. J., Thurgate S. et al., Appl. Phys. Lett. (1999 in press)
373. Robertson J. and Warren W. L., MRS Proc. **361**, 117 (1994)
374. Smyth D. M., Ann. Rev. Mat. Sci. **15**, 329 (1985)
375. Pintilie L., Alexe M., Pintilie I., and Boierasu I., Ferroelec. **201**, 217 (1997).
376. Kudzin A. Yu., Panchenko T. V., and Yudin S. P., Sov. Phys. Sol. St. **16**, 1589 (1974)
377. Pintilie L., Pintilie I., Petre D., Botila T., and Alexe M., Appl. Phys. A (in press 1999); Pintilie L., Alexe M., Pignolet A., and Hesse D., J. Phys. IV France **8**, 101 (1998)
378. Alexe M., Pintilie L., Pintilie I., Pignolet A., Senz S., and Hesse D., MRS Proc. **433**, 425 (1996); Appl. Phys. Lett. **69**, 1571 (1996)
379. Pintilie L. and Alexe M., J. Eur. Ceram. Soc. 19, 1485 (1999)
380. Hartmann A. J. et al., Integ. Ferroelec. (1998)
381. Melnick B. M., Abrokwah J., Hallmark J., and Ooms B., Integ. Ferroelec. **15**, 221 (1997)

Chapter 11

382. Amanuma K., Hase T., and Miyasaka Y., Jpn. J. Appl. Phys. **33**, 5211 (1994)
383. De Keijser M. et al., *Ferroelectric Thin Films*, eds. C. Paz de Araujo, J. F. Scott, and G. W. Taylor (Gordon & Breach, New York, 1996) p.485

384. Fujii E. et al., Int. Elec. Dev. Meeting (IEDM) (1992) p.267
385. Auciello O., Dat R. and Ramesh R., *Ferroelectric Thin Films*, eds. C. Araujo, J. F. Scott, and G. W. Taylor (Gordon & Breach, New York, 1996) p.525
386. Paz de Araujo C. A., McMillan L. D., and Scott J. F., MRS Proc. **230**, 277 (1991)
387. McMillan L. D., et al., Integ. Ferroelec. **2**, 351 (1992)
388. Isobe C. et al., International Symposium of Integrated Ferroelectrics, Santa-Fe, New Mexico, March 1997; Integ. Ferroelec. **14**, 95 (1997)
389. Ami T., Hironaka K. et al., MRS Proc. **415**, 195 (1996)
390. Tao W., Desu S. B., and Li T. K., Mater. Lett. **23**, 177 (1995)
391. Li T. K. et al., Appl. Phys. Lett. **68**, 616 (1996); MRS Proc. **415**, 189 (1996)
392. Kim D. C., Jo W., Lee H. M., and Kim K. Y., Integ. Ferroelec. **18**, 137 (1997)
393. Gutleben C. D., Appl. Phys. Lett. **71**, 3444 (1997); McKee R. et al., Phys. Rev. Lett. **81**, 3017 (1998); Alexe M., Physics World **12**, 21 (1999)
394. Schwee L., private communication (1986)
395. Fleddermann C., Integ. Ferroelec. **5**, 29 (1994); Uchida H., Soyama N., Kageyama K., Ogi K., and Paz de Araujo C. A., FMA-13 Conf., Kyoto (1996), p.193; Faure S. P., Chaux O., and Gaucher P., J. Phys. IV France **8**, 60 (1998)
396. Turova N. Ya.and Yanovskaya M. I., *Ferroelectric Thin Films*, eds. Paz de Araujo C., Scott J. F. and Taylor G. W. (Gordon & Breach, New York, 1996) p.233; Hanold R. C. F., U. S. Patent #4,081,857 (1975)
397. Kato K., Integ. Ferroelec. **22**, 13 (1998)
398. Dey S. K., *Ferroelectric Thin Films*, eds. Paz de Araujo C., Scott J. F. and Taylor G. W. (Gordon & Breach, New York, 1996) p.329
399. Auciello O., Kraus A. R., and Gifford K. D., *Ferroelectric Thin Films*, eds. Paz de Araujo C., Scott J. F. and Taylor G. W. (Gordon & Breach, New York, 1996) p.393
400. Krupanidhi S. B., Hu H., and Fox G. R., *Ferroelectric Thin Films*, eds. Paz de Araujo C., Scott J. F. and Taylor G. W. (Gordon & Breach, New York, 1996) p.427
401. Iijima K., et al., J. Appl. Phys. **60**, 361 (1986); **60**, 2914 (1986)
402. Adachi H., Mitsuyu T., Yamazaki O., and Wasa K., J. Appl. Phys. **60**, 736 (1986)
403. Sreenivas K. et al., J. Appl. Phys. **61**, 411 (1987); Appl. Phys. Lett. **52**, 709 (1988); Thin Solid Films **172**, 251 (1989)
404. Matsubara M. et al., MRS Proc. **200**, 243 (1990); **243**, 281 (1992)
405. Amanuma K., Hase T., and Miyasaka Y., Jpn. J. Appl. Phys. **33**, 5211 (1994)
406. Lee J. K., Song T. K., and Jung H. Y., Integ. Ferroelec. **15**, 115 (1997)
407. Paek S. H., Park C. S., Won J. H., and Lee K. S., MRS Proc. **443**, 33 (1996)
408. Yamamichi S. et al., IEEE IEDM (1995) p.119
409. Yang C.-H., Park S.-S., and Yoon S.-G., Integ. Ferroelec. **18**, 377 (1997)
410. Kalkur T. S. et al., Integ. Ferroelec. **1**, 327 (1992); MRS Proc. **230**, 315 (1992); Integ. Ferroelec. **5**, 177 (1994)
411. de Keijser M., Dormans G. J. M., and Larsen P. K., *Ferroelectric Thin Films*, eds. Paz de Araujo C., Scott J. F. and Taylor G. W. (Gordon & Breach, New York, 1996) p.485
412. Burn I., US Patent #4,120,677 (1978)
413. Sakashita et al., Appl. Phys. Lett. **73**, 7857 (1993); J. Appl. Phys. **57**, 2431 (1990)
414. Mazure C., IEEE ISAF (Lausanne, Sept. 1998)
415. Gardiner R., Brown D. W., Kirlin S. P., and Rheingold A. L., Chem. Mater. **3**, 1053 (1991); Lesaicherre P.-Y., Ferroelec. Newsletter **2**, 15 (1994); Oyo Butsuri Gakkai **41** (1994)

416. Kirlin P., Bilodeau S., and Van Buskirk P., Integ. Ferroelec. **7**, 307 (1995); Yuuki A. et al., IEEE IEDM (1995) p.115
417. Eguchi K. and Kiyotoshi M., Integ. Ferroelec. **14**, 33 (1997)
418. Kim N. G., Yoon S. G., Lee W. J., and Kim H. G., Integ. Ferroelec. **14**, 105 (1997)
419. No K. et al., MRS Proc. **443**, 9 (1996)
420. Kato Y. et al., MRS Proc. **443**, 3 (1996); Neumayer D. A. et al., Integ. Ferroelec. **18**, 297 (1997)
421. Kessler V. G. et al., Chem. Mater. **6**, 2336 (1994); Lindner J. et al., J. Phys. IV France **8**, 247 (1998)
422. Kaiser D. L., Vaudin M. D., et al., J. Mater. Res. (1999 in press); Dedyk A. I. et al., J. Phys. IV France **8**, 217 (1998).Watanabe K. et al., Appl. Phys. Lett. **73**, 126 (1998); Atsuki T., Soyama N., Yonezawa T., and Ogi K., Jpn. J. Appl. Phys. **34**, 5096 (1995); Noguchi T., Hase T., and Miyasaka Y., Jpn. J. Appl. Phys. **35**, 4900 (1996)
423. McMillan L. D., Araujo C. A., Roberts T., Cuchiaro J., Scott M. C., and Scott J. F., Integ. Ferroelec. **1**, 351 (1992); Solayappan et al., Integ. Ferroelec. **18**, 127 (1997)
424. Huffman M., Integ. Ferroelec. **10**, 39 (1995)
425. Solayappan N. et al., Integ. Ferroelec **14**, 237 (1997)
426. Scott J. F., Ferroelec. Rev. **1**, 1 (1998)
427. Bai G.-R. et al., Integ. Ferroelec. **21**, 291 (1998)
428. Takemura K. et al., Appl. Phys. Lett. **64**, 2967 (1994); Jpn. J. Appl. Phys. **34**, 5224 (1995)
429. Jia Q. X. et al., Integ. Ferroelec. **19**, 111 (1998)
430. Dey S. K. and Alluri P. V., MRS Bull. **21**, 44 (1996)
431. Auciello O., Dat R., and Ramesh R , *Ferroelectric Thin Films*, eds. Paz de Araujo C., Scott J. F. and Taylor G. W. (Gordon & Breach, New York, 1996) p.525
432. Shimizu M., Katayama T., Shiosaki T. and Kawabata A., ISAF Proc. (IEEE, New York, 1990) p.669; Petuskey W. T., Richardson D. A., and Dey S. K., Integ. Ferroelec. **2**, 269 (1992); de Keijser M., Dormans G. J. H., and Larsen P. K., *Ferroelectric Thin Films*, eds. Paz de Araujo C. A., Scott J. F., and Taylor G. W. (Gordon & Breach, New York, 1996) p.485
433. Chrisey D. B. and Huber G. K., *Pulsed Laser Deposition of Thin Films* (Wiley, New York, 1994); Maher G. H., US Patent #4,266,265 (1979)
434. Roy D., Peng J. and Krupanidhi S. B., Appl. Phys. Lett. **60**, 2478 (1992)
435. Ramesh R. et al., Science **252**, 944 (1991)
436. Lin H. et al., Integ. Ferroelec. **5**, 197 (1994)
437. Lee J., Ramesh R. et al., Integ. Ferroelec. **5**, 145 (1994)
438. Suga M. et al., Integ. Ferroelec. **18**, 389 (1997)
439. Pignolet A. et al., Integ. Ferroelec. **22**, 485 (1998); Integ. Ferroelec. **24** (in press 1999); MRS Proc. **433**, 125 (1996); J. Phys. IV France **8**, 251 (1998); Matsumuro Y., Sugiyama H., Noda M., and Okuyama M. (in press)
440. Fujii E. et al., Int. Elec. Dev. Meeting (IEDM) (1992) p.267
441. Noma A. and Ueda D., Integ. Ferroelec. **15**, 69 (1997)
442. Gutleben C. D., MRS Proc. **433**, 109 (1996); MRS Proc. (1999 in press)
443. McKee R. et al., Phys. Rev. Lett. **81**, 3017 (1998); Alexe M., Physics World **12**, 21 (1999)

Chapter 12

444. Rabson T. A., Rost T. A., and Lin H., Integ. Ferroelec. **6**, 15 (1995)
445. Smith E. B., Lin H., Rost T. A., and Rabson T., Integ. Ferroelec. **3**, 85 (1993)
446. Kalkur T. S., Jacobs B. and Argos G., Integ. Ferroelec. **5**, 177 (1994)
447. Lin M. and Kalkur T. S., Integ. Ferroelec. **14**, 247 (1997)
448. Kalkur T. S., Kwor R. Y., Levenson L. and Kamerdiner L., Integ. Ferroelec. **1**, 327 (1992)
449. Sinharoy S. et al., IEEE Trans. Ultrason. Freq. **38**, 663 (1991); J. Vac. Sci. Technol. **A9**, 409 (1991); Integ. Ferroelec. **1**, 377 (1992)
450. Lampe D. R., Adams D. A., Sinharoy S., and Buhay H., Integ. Ferroelec. **3**, 121 (1993)
451. Aizawa K., Ichiki T. and Ishiwara H., MRS Proc. **310**, 313 (1993)
452. McMillan L. D., reproduced in Scott J. F., Ferroelec. Rev. **1**, 1 (1998)
453. Kalkur T. S., Integ. Ferroelec. **3**, 351 (1993)
454. Autran J. L. et al., Suppl. *Le Vide: Science, Technique, et Applications* **275**, 44 (1995) [Proc. 2nd Int. Conf. Space Charge in Solid Dielectrics, Antibes, 1995]
455. Watanabe Y., Tamamura M., and Matsumoto Y., Jpn. J. Appl. Phys. **35**, 1564 (1996)
456. Ishiwara H., Jpn. J. Appl. Phys. **32**, 442 (1993)
457. Ishiwara H., Shimamura T., and Tokumitsu E., Jpn. J. Appl. Phys. **36**, 1655 (1997)
458. Alexe M., Pignolet A., Senz S. and Hesse D., Ferroelec. **201**, 157 (1997) achieve a memory window of 3.35 V with bismuth titanate sol–gel films; for other ferroelectric gate materials, see Proc. SSDM, Jpn. J. Appl. Phys. **38** (1999)
459. McKee R. et al., Phys. Rev. Lett. **81**, 3017 (1998)
460. Alexe M., Physics World **12**, 21 (1999)
461. Alexe M., Appl. Phys. Lett. **72**, 2283 (1998)
462. Gaucher P., Eichner D., Hector J. and Von Munch W., J. Phys. IV France **8**, 235 (1998)
463. Alexe M. et al., J. Phys. IV France **8**, 239 (1998)

Chapter 13

464. Yandrofski R., et al., US Patent #5,472,935 (1995)
465. Herman A. M. et al., Bull. Am. Phys. Soc. **38**, 689 (1993)
466. Hermann A. M. et al., J. Superconduct. **7**, 463 (1994)
467. Barnes F. S., et al., Integ. Ferroelec. **8**, 171 (1995)
468. Galt D., Price J. C., Beall J. A., and Ono R. H., Appl. Phys. Lett. **63**, 3078 (1993)
469. Jackson J. D., *Classical Electrodynamics* (Wiley, New York, 1962) p.264
470. Scott J. F. et al., Integ. Ferroelec. **6**, 189 (1995); Galt D., Price J. C., and Ono R. H., IEEE MTT-S Int. Micowave Symp. Digest (1993) p.1421; Zafar S. et al., Appl. Phys. Lett. **72**, 2820 (1998)
471. Theis C. D. et al., Thin Solid Films **325**, 107 (1998); Appl. Phys. Lett. **72**, 2817 (1998); Kaiser D. L. et al., J. Mater. Res. (1999, in press)
472. Missert N. et al., IEEE Trans. Appl. Superconduct. **13**, 1741 (1993)
473. Integrated Ferroelectrics, Vol. 22 (1998)
474. Jack L., US Patent #5,070,241 (1991)
475. Babbitt R. W. et al., US Patent #5,212,463 (1993)

476. Collier D. C., Integ. Ferroelec. **4**, 113 (1994)
477. Kain A. et al., Integ. Ferroelec. **8**, 45 (1995)
478. Jackson C. M. et al., Integ. Ferroelec. **4**, 121 (1994)
479. Wilber W., et al., Integ. Ferroelec. **19**, 149 (1998)
480. Sengupta L. C. et al., Integ. Ferroelec. **8**, 77 (1995)
481. Babbitt R. W., Koscica T. E., and Drach W. C., Microwave J. **35**, 63 (1992)
482. Babbitt R. W. et al., Integ. Ferroelec. **8**, 65 (1995)
483. Tokunaga M., J. Phys. Soc. Jpn. **56**, 1653 (1987)
484. Tokunaga M., J. Phys. Soc. Jpn. **57**, 4275 (1988)
485. The original work on pure cadmium titanate [Smolensky A., JETP **20**, 137 (1958)] and on lead pyrochlore [Hulm A., Phys. Rev. **92**, 504 (1953)] did not demonstrate switching, but on the basis of recent work on mixed crystals of form $Ca_{2-2x}Pb_{2x}Nb_2O_7$ and $Ca_xCd_{1-x}TiO_3$, it is clear that the pure materials are ferroelectric at and below the stated temperatures
486. Petrov P. K . et al., J. Appl. Phys. **84**, 3134 (1998)
487. Kozyrev A. B. et al., J. Appl. Phys. **84**, 3325 (1998)

Chapter 14

488. Alexe M., Kastner G., Hesse D., and Goesele U., Appl. Phys. Lett. **70**, 3416 (1997); J. Phys. IV France **8**, 239 (1998); Mater. Chem. and Phys. **55**, 55 (1998); J. Korean Phys. Soc. **32**, S1618 (1998)
489. Alexe M., Scott J. F., Pignolet A., Hesse D., and Goesele U., Integ. Ferroelec. **19**, 95 (1998); Takeishi T. and Whatmore R. W., J. Phys. IV (France) **8**, 57 (1998)
490. Budd K. D., Dey S. K., and Payne D. A., Brit. Ceram. Proc. **36**, 107 (1985)
491. Stengl T., Ahn K.-Y., and Goesele U., Jpn. J. Appl. Phys. **27**, 236 (1988)
492. Maszara W. P., Goetz G., Caviglia A., and McKitterick J. B., J. Appl. Phys. **64**, 4943 (1988)
493. Watton R., Ferroelec. **91**, 87 (1989); **133**, 5 (1992); Integ. Ferroelec. **4**, 175 (1994)
494. Shorrocks N. M., Patel A., and Whatmore R. W., Ferroelec. **134**, 343 (1992)

Chapter 15

495. Belayev L. M. and Bendrikova G. G., Fiz. Tverd. Tela **6**, 645 (1964)
496. Rosenblum B., Braunlich P., and Carrico J., Appl. Phys. Lett. **25**, 17 (1974)
497. Jona F. and Shirane G., *Ferroelectric Crystals* (Pergamon, Oxford, 1962); Fatuzzo E. and Merz W. J. (North-Holland, Amsterdam, 1967)
498. Gundel H. et al., Appl. Phys. Lett. **54**, 2071 (1989); J. Appl. Phys. **69**, 975 (1991)
499. Gundel H., Integ. Ferroelec. **2**, 207 (1992)
500. Biedrzycki K., Phys. Stat. Sol. **93**, 503 (1986)
501. Rosenman G. I., Ferroelec. **141**, 421 (1993)
502. Rozenmann G. I. et al., JETP Lett. **39**, 477 (1984)
503. Gundel H., private communication
504. Benedek G. et al., J. Appl. Phys. **83**, 2766 (1998)

Chapter 16

505. Dimos D. et al., MRS Proc. **310**, 87 (1993); J. Appl. Phys. **76**, 4305 (1994)
506. Brody P. S., Ferroelec. **38**, 939 (1981)
507. Brody P. S. and Rod B. J., Integ. Ferroelec. **2**, 1 (1992)
508. Ignatiev A. et al., Integ. Ferroelec. **10**, 327 (1995)
509. Wang F. et al., Opt. Lett. **18**, 615 (1993)
510. Wang F. and Haertling G. H., Appl. Phys. Lett. **63**, 1730 (1993)
511. Wang F. and Haertling G. H., Optics and Photonics News **48** (1993)
512. Warren W. L. et al., MRS Bull. **21**, 40 (1996); Appl. Phys. Lett. **65**, 1018 (1994)
513. Wu S.-Y., Ferroelec. **11**, 379 (1975)
514. Pintilie L., Alexe M., Pignolet A., and Hesse D., Appl. Phys. Lett. **73**, 342 (1998)
515. Alexe M. and Pintilie L., Infrared Phys. & Technol. **36**, 949 (1995); Lang S. B. and Alexe M. (1999 in press)

Chapter 17

516. Alexe M., ISIF, Colorado Springs (March 1999), Integ. Ferroelec. (in press)
517. Okamura S., Mori K., Tsukamoto T., and Shiosaki T., Integ. Ferroelec. **18**, 311 (1997)
518. Scott J. F. et al., Integ. Ferroelec. **21**, 1 (1998)
519. Alexe M. et al., Appl. Phys. Lett. **73**, 1592 (1998); Switzer J. A., Shumsky M. G., and Bohannan E. W., Science **284**, 293 (1999)
520. Yu B., Zhu C., and Gan F., J. Appl. Phys. **82**, 4532 (1997)
521. Taylor G. I., Proc. Roy. Soc. (London) **A280**, 383 (1964); Turnbull D., J. Appl. Phys. **23**, 1022 (1950)
522. Wakayama Y. and Tanaka S.-I., Proc. Nano'98, Stockholm (June 16, 1998) p.49; Nanostructured Materials **12** 13 (1999)
523. Moore J. T. et al., Appl. Phys. Lett. **72**, 1840 (1998); **72**, 1254 (1998); Burmeister F. et al., Adv. Mater. **10**, 495 (1998)
524. Watanabe K. et al., Japanese patent application # H08-261500
525. Zafar S. et al., J. Appl. Phys. **82**, 4469 (1997)
526. Watanabe K., Scott J. F., Hartmann A. J.; Appl. Phys. (in press, 1999); Hartner W. et al., Integ. Ferroelec. **22**, 23 (1998)
527. Watanabe K., Hartmann A. J., and Scott J. F., Appl. Phys. (in press 1999)
528. Isobe C. et al., Integ. Ferroelec. **14**, 95 (1997)
529. Alexe M., Goesele U. et al., submitted to Science (1999)
530. Amanuma K. and Kunio T., Jpn. J. Appl. Phys. **35**, 5229 (1996)
531. Scott J. F., *Thin Film Ferroelectric Materials and Devices*, ed. R. Ramesh (Kluwer Academic, Dordrecht, 1997) p.115
532. Sushkov O., private communication

Chapter 18

533. Park B. H., Kong B. S., Bu S. D., Noh T. W., Lee J., and Jo W., Nature **401**, 682 (1999)
534. Sanchez L. E., Dion D. T., Wu S.-Y., and Naik I. K., Ferroelec. **116**, 1 (1991)
535. Dietz G. W., Schumacher M. and Waser R., *Science and Technology of Electroceramic Thin Films*, eds. Waser R. M. and Auciello O. (Kluwer, Dordrecht, 1995) p.269. For an earlier study of fast current transients in dP/dt see Benedetto J. M., Moore R. A., and McLean, F. B., Integ. Ferroelec. **1**, 195 (1992). Cottrell desorption is discussed by Richert H. and Steiner R., Z. Phys. Chem. **49**, 127 (1966)
536. Scott J. F., Habbal F., and Zvirgzds J. A., J. Chem. Phys. **72**, 2760 (1980)
537. Zafar S. et al., Appl. Phys. Lett. **73**, 175 (1998)
538. Williams R., J. Phys. Chem. Sol. **22**, 129 (1961); Phys. Rev. **125**, 850 (1962); Many A., J. Phys. Chem. Sol. **26**, 575 (1965); Boer K. W. and Kummel U., Z. Phys. Chem. **200**, 180 (1952); Z. Naturforsch. **9A**, 117 (1954); **16A**, 678 (1958); Ann. Phys. (Leipzig) **14**, 341 (1954); **20**, 303 (1957)
539. Wittels M. C. and Sheril F. A., J. Appl. Phys. **28**, 606 (1957); Lefkowitz I. and Mitsui T., J. Appl. Phys. **30**, 269 (1959)
540. Chynoweth A. G., Phys. Rev. **102**, 705 (1956)
541. Fatuzzo E., Helv. Phys. Acta **33**, 502 (1960)
542. Boutin H., Frazer, and Jona F., J. Chem. Phys. Sol. **24**, 1341 (1963)
543. Scott J. F., Paz de Araujo C. A., and McMillan L. D., Ferroelec. **116**, 107 (1991); Scott J. F. et al., J. Appl. Phys. **66**, 1444 (1989)
544. Nasby R. D., Schwank J. R., Rodgers M. S., and Miller S. L., Integ. Ferroelec. **2**, 91 (1992); Zheng L., Lin C., and Rao K. V., J. Phys. IV France **8**, 265 (1998)
545. Taylor G. W., Ferroelec. **18**, 17 (1978)
546. Moll J. L. and Tarui Y., IEEE Trans. Electron Dev. **10**, 328 (1963)
547. Zuleeg R. and Wieder H. H., Sol. St. Electron. **9**, 657 (1966)
548. Heyman P. M. and Heilmeyer G. H., Proc. IEEE **54**, 842 (1966)
549. Perlman S. S. and Ludwig K. H., IEEE Trans. Electron Dev **14**, 816 (1967)
550. Teather G. G. and Young L., Sol. St. Electron. **11**, 527 (1968)

Index

Druck: Strauss Offsetdruck, Mörlenbach
Verarbeitung: Schäffer, Grünstadt